GERMAN VISOR CAPS

OF THE SECOND WORLD WAR

GERMAN VISOR CAPS

OF THE SECOND WORLD WAR

HEER

LUFTWAFFE

KRIEGSMARINE

WAFFEN-SS

GUILHEM TOURATIER AND
LAURENT CHARBONNEAU

Translated from the French by Omicron Language Solutions, LLC.

This book was originally published under the title,
Les Casquettes Allemandes de la Seconde Guerre Mondiale,
by REGI'ARM, Paris, France.

Copyright © 2013 by Schiffer Publishing, Ltd.
Library of Congress Control Number: 2013946084

Printed in China.
ISBN: 978-0-7643-4458-9

We are interested in hearing from authors with book ideas on related topics.

Published by Schiffer Publishing Ltd.
4880 Lower Valley Road
Atglen, PA 19310
Phone: (610) 593-1777
FAX: (610) 593-2002
E-mail: Info@schifferbooks.com.
Visit our web site at: **www.schifferbooks.com**
Please write for a free catalog.
This book may be purchased from the publisher.
Try your bookstore first.

CONTENTS

Acknowledgements:

Ludwig Bauer, Frederic Coune, Manfred Zauner, Hervé Bertin, Didier Zawis, Philippe Rio & Richie C., Cyril Monties, Dr Dupont, Sina Eberl, Frieder Ortlepp, Ambleteuse Historical Museum of the Second World War (Pas de Calais), Xavier Aiolfi, the Touratier, Hofmann and Hufnagl families, Max Lötsch and everyone, near and far, who helped with this project.

BIBLIOGRAPHY

■ **Eberhard Hettler** (Hauptmann im RLM), *Uniformen der deutschen Wehrmacht,* Uniformen-Markt Verlag Otto Dietrich, Berlin SW 68, 1939.
■ **Jill Halcomb and Wilhelm P.B.R. Saris with Otto Spronk,** *Headgear of Hitler´s Germany (Vol. 1 & 2),* R. James Bender Publishing, San Jose, 1989.
■ **Gary Wikins,** *The Collector´s Guide to Cloth Third Reich Military Headgear,* Schiffer Publishing Ltd., Atglen.
■ **John R. Angolia & Adolf Schlicht,** *Uniforms & Traditions of the Luftwaffe (1-3),* R. James Bender Publishing, 1996.
■ **John R. Angolia & Adolf Schlicht,** *Uniforms & Traditions of the German Army (1-3),* R. James Bender Publishing, 1984.
■ **John R. Angolia & Adolf Schlicht,** *Die Kriegsmarine – Uniforms and Traditions (1-3),* R. James Bender Publishing, 1991.
■ *Uniformen und Abzeichen, herausgegeben von Obltn.* a. **D. Jul. M. Ruhl und Major a. D. A. Sußmann Neu bearbeitet von C. Starke,** Verlag Moritz Ruhl, Leipzig O 27, vers 1941-42.
■ **Jean Eparvier – Jean-Louis Babelay,** *A Paris, sous la botte des nazis,* Editions Raymond Schall, novembre 1944.
■ **Pat Moran & Jon Maguire,** *German Headgear in World War II - Army, Luftwaffe & Kriegsmarine, A Photographic Study of Hats and Helmets* – Schiffer Publishing Ltd., Atglen, 1997.
■ **Michael D. Beaver,** *Uniforms of the Waffen-SS,* Schiffer Publishing Ltd., Atglen, 2002.
■ **Richard de Filippi,** *Les coiffures militaires du Troisième Reich,* Jacques Grancher, 1980.
■ *Ordonnances de l'armée allemande : Allgemeine Heeresmitteilungen,* Heeres-Verordnungsblätter (fonds allemands du S.H.D. - Vincennes-, Militärarchiv de Fribourg).
■ *Orden und Ehrenzeichen, Abschnitt I,* Merkblatt 15/5, 1943 (S.H.D. - F.A. 13756).
■ *Zentner/Bedürftig, Das grosse Lexikon des zweiten Weltkriegs,* Südwest, München, 1998.
■ **G. Tessin,** *Verbände und Truppen der deutschen Wehrmacht und Waffen-SS 1939/45,* Biblio Verlag, Osnabrück, 1975.
■ **E. Lefèvre,** *La Wehrmacht (uniformes et insignes),* Grancher, 1986.
■ *Germania Insignia of World War II,* edited by **Chris Bishop and Adam Warner,** Grange Books, 2002.
■ **A. Mollo,** *German Uniforms of World War Two,* Hippocrene Books, New-York, 1976.

"You can tell a soldier by his uniform," was Emperor Napoleon I's favorite phrase, and as he was almost fanatical about the smallest details of his army's uniform, he knew what he was talking about.
In any army throughout the world, the hat worn by the soldier is the first thing we notice; it heightens his power and gives him an air of authority, just as in the Middle Ages a helmet wasn't just for protection, it was almost a second face.

If a hat defines military prowess, it also gives precious information about the man wearing it.
In fact, a cap becomes identified with the person who wore it, in his choice of its form, folds or its height. It therefore reflects an image and gives clues to anyone who wants to understand its former owner's personality.

Finally, if the diverse holes and general wear and tear are stigmas bearing mute witness to the ferocity of war, some types of cap are symbols of certain battles or certain events.
Thus the Battle of the Atlantic can be resumed in the white caps of the Navy's company-grade officers, which speak to us louder than words especially when they are stained with motor oil.
All these items have (luckily) long since left behind them the propaganda and sufferance of war and are now a part of museum expositions and collector's showcases, where they are a tangible trace of the history of mankind.

Lastly, I would like to pay tribute to the entire team of the "Uniforms" magazine who with
the help of our catalogues going back over thirty-five years, have given history and collectors alike a superb reference book.

Wolfgang Hermann
Founder, Hermann Historica, OHG

INTRODUCTION

Consciously or not, the Schirmmütze remains inseparably linked to the image of German officers and soldiers of the Second World War. In truth, this elegant cap, with its well thought-out design, has stayed in the minds of the people who were able to observe them during that epoch and who have passed on a fantasized, sometimes erroneous vision, often mixed up with contradictory emotions.

However, all these reactions aren't unfounded as the Schirmmütze really was, as we will see, the traditional headwear par excellence of the German Wehrmacht.

▲ Three Waffen-SS officers posing at Hitler's Berghof in May 1941. We can clearly see their caps, which had a Totenkopf affixed to the black velvet band. From left to right: Fritz Darges (Hitler's aide-de-camp), Fritz Klingenberg and Laackmann. Klingenberg wears the Knight's Cross and the belt of an officer of the Waffen-SS. Laackmann wears the Wound Badge in Black and the Iron Cross 1st Class. His Totenkopf collar tabs are embroidered in silvered wire.

▲ A pile of Wehrmacht officers' caps at a theater – witness to the enthusiasm the conquerors had for shows in occupied Paris.

▶ During the occupation, "Old England," a famous Parisian store, sold uniforms to the German Wehrmacht. (Collection XA)

HEER

The Heer was the name given to the land forces of the German Wehrmacht between 1933 and 1945. It was during this period that the Schirmmütze, with its harmonious shape and insignia, remained fixed in people's memories. It became an indispensible attribute that represented German officers of the Second World War, and through these pages we will see that it was also intended for NCOs and the troops. The efficient impact it had came not only from its simplicity but also from its numerous variants that constituted interesting information about those who wore them, and who we will talk about throughout this book.

▶ German cap worn by NCOs at the end of World War I.

NON-COMMISSIONED OFFICERS AND TROOPS

▲ **Feldmütze of the Reichswehr,** worn here with the first tunic bearing the national emblem, the eagle with swastika, of the new Heer. The tunic of this Obergefreiter is a 1933 model, as can be seen by the collar tabs and the visible stitching above the breast pockets. The "type 1" eagle above the right breast pocket is woven with white cotton on a light field-grey background, the collar tabs resemble a tilted Roman numeral II, the pointed shoulderboards are plain, without rank insignia and are field-grey.

▲ **Dienstmütze worn by NCOs in Saxon units:** field-grey wool with reseda band and piping, metal insignia and ersatz patent black material visor; the lining is violet.

▲ **Feldmütze worn by Reichswehr Troops and NCOs:** greenish-grey wool, band and reseda green piping, metal cockade. Lining is white cotton with several stamps and marks denoting Infantry Regiment 16. This type of cap is very rare.

▲ Gathered around a unit Warrant Officer (Spiess) these NCOs of Artillerie-Regiment 15 pose in front of a barracks in Frankfurt. They all wear "Teller form" Schirmmützen. This photo was taken before the war, around 1938. On the right of the Spiess is a standard bearer.

▲ Two company Warrant Officers accompany a Leutnant, wearing an old model Feldmütze, on a visit to newly conquered Paris. The Spiess at right wears a "Teller form" cap. The dark color of their cap piping means that these men could belong to a Pionier unit.

▲ This image of a Reichswehr soldier gives us a glimpse of his Dienstmütze with its eagle surrounded by a wreath of oak leaves, and the cockade in army colors; typical of the Weimar Republic era.

▶ Maximillian Hofmann, a young recruit in the Panzerjäger, in dress uniform, around 1940. He wears a Waffenrock and a superb troop cap that, according to the rules of the new Heer, should be on the same level as the eyebrows (as seen here).

HEER :
CAP CHRONOLOGY

It was in **1921** that standard issue caps were adopted by the Reichswehr for all military personnel and worn when the field cap or helmet were not appropriate. A new addition since World War I was the piping which was no longer field-grey, a leather chinstrap, though the band remained field-grey.

From **1926**, the chinstrap consisted of three parts with rectangular buckles and was adjustable. In fact, some of the buckles were oval shaped.

1927: Officers' caps were given a silver braid; generals a gold braid (or gold piping).

1930: The upper piping on the general staff (officials) caps was changed to green.

1931: The form of the standard issue caps was changed: the front was raised, the peak's slope was more pronounced; inside a wire hoop was added, and the lining was orange-brown, with a diamond shaped celluloid sweatshield, the upper part of which was rounded off.

The general look of both officers' caps and those privately purchased evolved and changed from a flat form (Teller form, plate-shaped, sometimes called "Reichswehr shaped") to a saddleback form.

1934: From this date on, standard caps were called Schirmmütze (peaked cap). They were intended for NCOs and the troops when wearing dress uniform or when being presented to superior officers. Senior NCOs were authorized to wear them with service uniform and undress fatigues. During peacetime, each man had two caps (with the number I or II inside) that had to last for a minimum of at least one year. In 1939 battledress was adopted and the Schirmmütze were left in the barracks; they were, however, obligatory when in dress uniform and were often privately purchased (this was possible from 1920 to 1942).

Since **1920**, standard issue caps had an orange-brown lining, rather similar to the color under the peak as well as the interior leather edge (the two ends of which joined on the side and not on the back. Above the lining there was often a piece running round the edge made of cork). The size was written under the celluloid sweatshield in ink; this shield was diamond-shaped, rounded off at one end (a paper label with the owner's name could also be slipped in a space provided for this purpose). On the inside leather edge the date and manufacturer's name were stamped. An allocation stamp was also sometimes added.

From **1939**, they were supposed to be camouflaged according to mobilization measures.

▲ Unteroffizier Bartholdi, unit commander of Panzerjäger Ersatz Kompanie 45, is shown here wearing a "Teller-form" cap. He also wears the Infantry Assault lanyard, the 1st October 1938 Medal with Prague Castle bar, and the Reich sports insignia. (Collection Hofmann/Lotsch)

▲ **Schirmmütze for Infantry Troops and NCOs,** made of field-grey serge, with metal insignia and white piping. Orange lining, the sweatband is made of ersatz material.

▶ **Promotional pocket mirror:** Friedrich Egg, maker of hats, caps, military effects. Lindau, Bodensee, Lake Constance, founded in 1702. (Private Collection)

Friedri Hüte / Mützer Lin

◀ **Schirmmütze for NCOs and Troops of the Kradschützen.** Field-grey cap with copper colored piping, metal insignia and black, varnished chinstrap. Gold-colored lining and manufacturer's stamp: "Aug. Müller - München."

◄ **Schirmmütze for Panzer NCOs and Troops.** Field-grey serge, dark green band, pink piping, metal insignia. Gold colored lining with "EREL Sonderklasse Extra" celluloid sweatshield.

► Photo of an infantry Gefreiter. Note on the photo: "2. Abmarsch nach Rußland, februar 43" ("Second tour in Russia, February 1943"). This Gefreiter has slightly altered the shape of his cap.

▲ **Cap for NCOs and Troops of the Nebeltruppe.** Field-grey gabardine, dark green band, maroon piping, metal insignia and varnished leather chinstrap. Cotton lining with leather sweatband.

▲ **Cap for NCOs and Troops of the Cavalry. Field-grey serge, green band, gold colored piping, and aluminum oak-leaf wreath. Lining gold colored rayon with celluloid, diamond-shaped sweatshield, and leather sweatband. The national emblem was removed after the war. Above: photo of the former owner.**

▲ **Feldwebel decorated with the Infantry Assault Badge. The shoulderboards have white piping, the unit buttons would seem to belong to the infantry, while the piping on the cap seems much darker; the insignia is of stamped metal, and the chinstrap is black leather.** (Collection Hervé Bertin)

▲ **Cap for NCOs and Troops of the Geheime Feldpolizei.**
Field-grey Serge, brown rather than dark green band, light blue piping and silver-metal insignia. The lining has a celluloid sweatshield stamped "Sonderklasse," and the sweatband is made of leather.

▲ Gefreiter of the Ludwigsburg Infantry (region of Stuttgart). The silver buttons on his shoulder boards give us an idea of his senior NCO status. His tunic and insignia, and collar tabs are piped white for infantry; sharp shoulderboards, denoting an early manufacture; he also has an Infantry Assault lanyard. His cap has the new insignia added in 1935. (Collection Hervé Bertin)

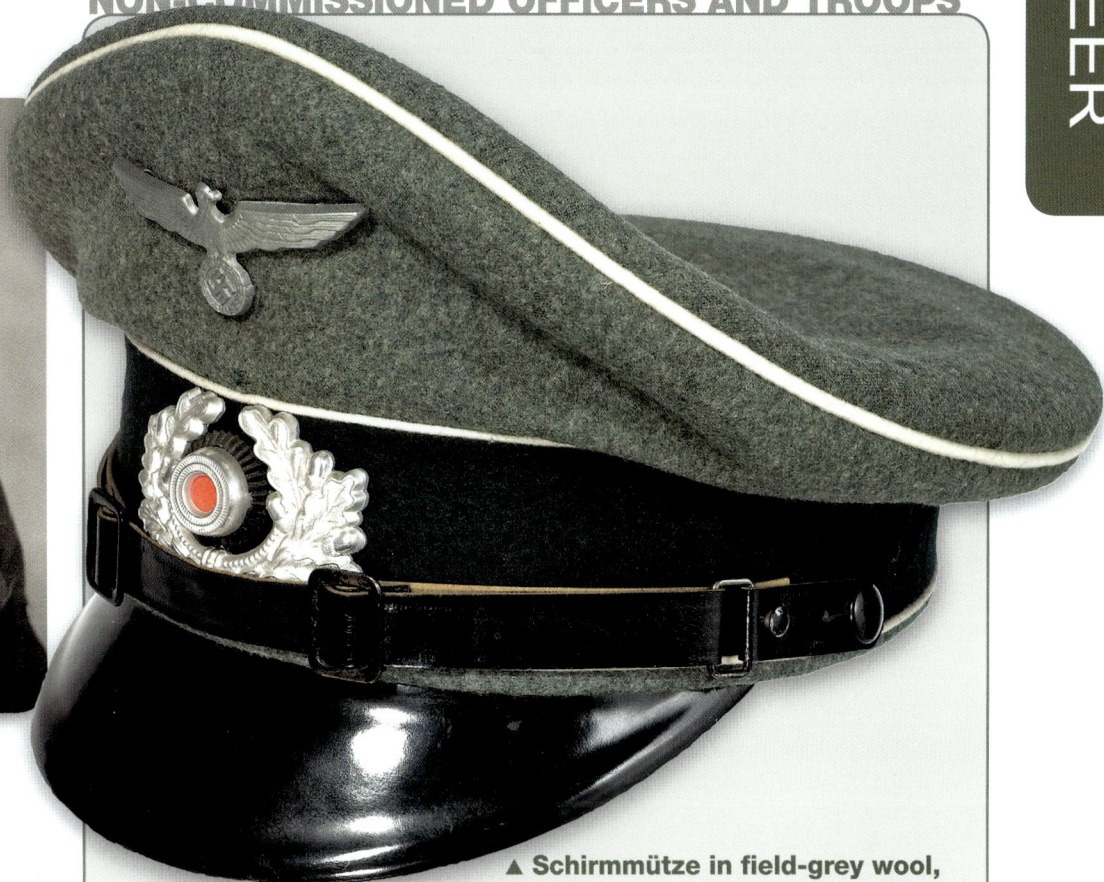

▲ Schirmmütze in field-grey wool, (infantry) with dark green band and white piping.

▲ Infantry Feldwebel. His cap has a troop/NCO leather chinstrap typical of SS caps.

▲ Cap for NCOs and Troops of Artillerie. Field-grey serge, dark green band, red piping, aluminum insignia, gold colored lining, celluloid sweatshield marked "Schellenberg Sonderklasse," and leather sweatband. Size 57.

▲ **Schirmmütze for NCOs and Troops of Artillerie:** Field-grey serge, dark green band, red piping, and aluminum insignia. Brown lining with celluloid sweatshield bearing the manufacturer's name "Peküro"; leather sweatband. Size 58.

▲ **Schirmmütze for Infantry NCOs and Troops:** Field-grey textile, dark green band, white piping, thin zinc insignia. Grey rayon lining, gold mark "Deutsches Erzeugins" (made in Germany), ersatz material sweatband. Size 57.

▶ **The Heer** is represented here through diverse characteristic elements. On the left we have a figure of an officer of the 21st Panzer Division, in a uniform such as he would have worn during the Battle of Normandy: M1933 Hauptmann's tunic modified in 1936, over which is a reversible camouflage smock (white interior) and an infantry cap (white piping) that our officer chose to wear during the campaign; in spite of its vulcanized fiber peak and silver-wired insignia, it resembles an Alter Art model from its shape and the lack of an officer's cord. On the right is an officer of the Sturmartillerie, wearing a serge field-grey wrap, color piped for an artillery division (red). Our officer has received the Iron Cross 1st Class, and the Wound Badge in Black; around his neck are rare Bakelite 6x30 binoculars. The cap is an authentic Alter Art model with all its characteristics: varnished leather peak, insignia woven in aluminum reserved for officers, lack of cord and orange canvas lining, the same as can be seen in standard issue caps. Also included is a Sturmgeschütz training model, an M42 steel helmet camouflaged brown and green, an M1924 stick grenade, and a regulation P38 pistol. Finally, a pair of infantry Major's shoulderboards, and a well made Schirmmütze for an infantry officer with silver-wired insignia completes the ensemble.
(Private collection. Photo Frederic Coune)

▲ Formal photograph of an infantryman wearing a Schirmmütze, around 1935.
(Collection Hervé Bertin)

NON-COMMISSIONED OFFICERS

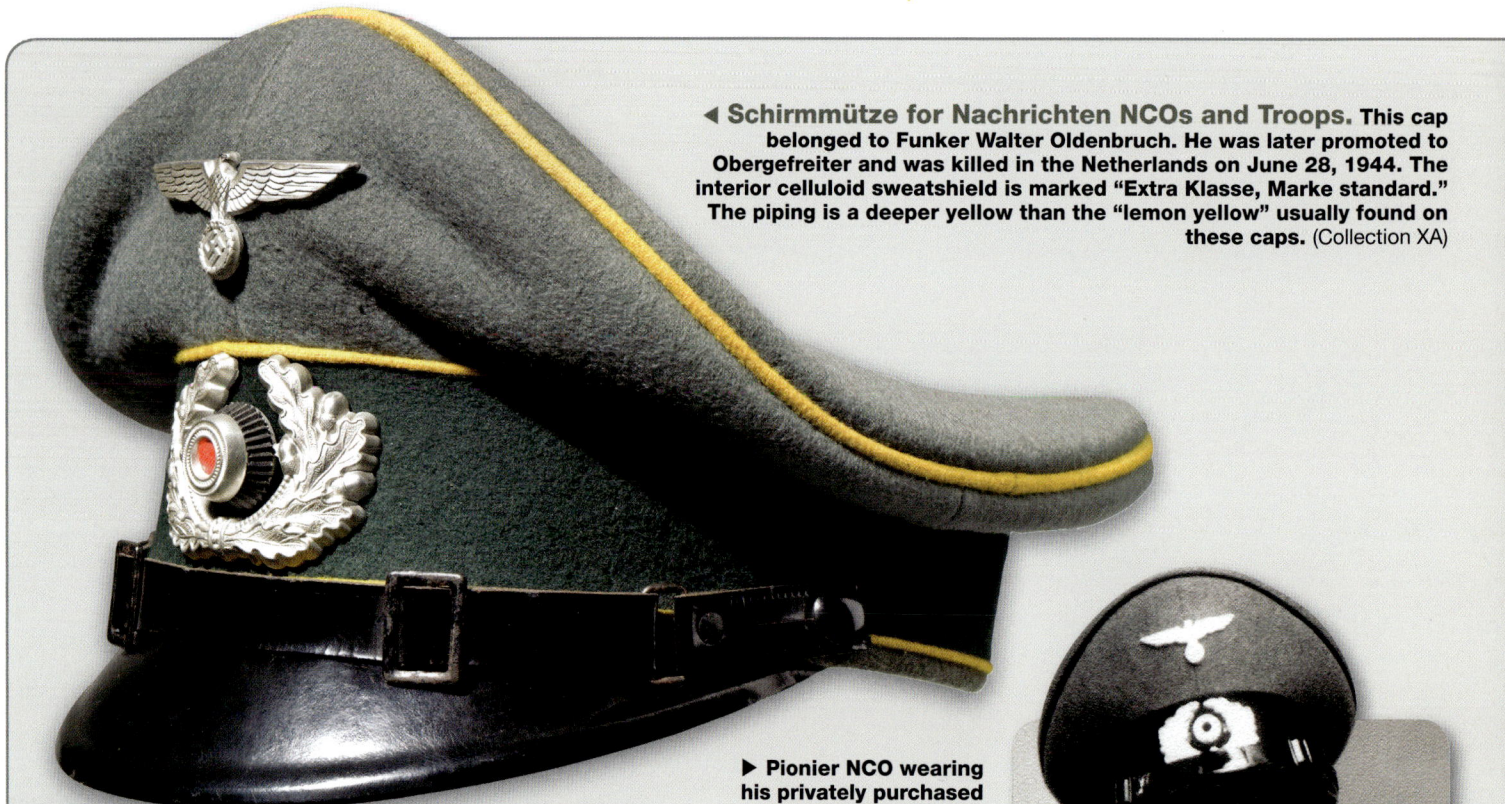

◄ **Schirmmütze for Nachrichten NCOs and Troops.** This cap belonged to Funker Walter Oldenbruch. He was later promoted to Obergefreiter and was killed in the Netherlands on June 28, 1944. The interior celluloid sweatshield is marked "Extra Klasse, Marke standard." The piping is a deeper yellow than the "lemon yellow" usually found on these caps. (Collection XA)

► **Pionier NCO** wearing his privately purchased Schirmmütze with 1935 insignia. Note the Anschluss ribbon, as well as what could be a Four Year Service ribbon.

◄ **Heer soldier** wearing a tunic piped in Pionier black. He wears a "Teller form" cap with metal insignia adopted in 1935.

► **Schirmmütze for NCOs and Troops** made of field-grey serge, Pionier black piping and silver metal insignia. Grey lining, sweatshield celluloid with the manufacturer's mark, "Hess Würzburg," leather sweatband.

▶ **Schirmmütze for NCOs and Troops of the 2.-.4./6 Cavalry Regiment,** field-grey gabardine with dark green band, gold-colored piping. Between the metal insignia is the gold metal Schwedt Eagle. The lining is brown.

▶ This mountain rifleman (photo 1939) is wearing a standard issue cap with the Edelweiss stamped in white metal. (Collection Franco Mesturini-Luca Soldati)

▶ **Superb Cap for NCOs of the Gebirgsjäger.** The Edelweiss insignia appeared in 1939 and is made of stamped white metal with a golden center. The interior bears the manufacturer's mark, "Gebrüder Hollofer, Salzburg, Getreidegasse 10, Frischluft D.R.G.M. 1419757." This type of cap is relatively rare, as the Gebirgjäger preferred the more traditional Bergmütze. Note the ventilation grid situated in the red centerpiece of the Reich cockade. (Collection XA)

▲ Detail of the Death's Head on a Schirmmütze worn by the cavalry. This model was inherited from the Hussars.

▲ **Brunswick Death's Head insignia affixed to a NCO Schirmmütze of the Cavalry.** At first reserved for the 4th Squadron of Reiter Regiment 13 (which became Cavalry Regiment 13 in 1936, and who kept the Brunswick Hussars Regiment 17 traditions), then in 1938, to the Squadron of the Cavalry Regiment 13 until its dissolution in 1939. The Brunswick insignia continued to be worn by the reconnaissance units and later in the war, by Cavalry Regiment 4. This cap is stamped "Sonderklasse" on the celluloid sweatshield. (Collection XA)

▲ Different NCO and troop chinstraps.

▲ Details of the manufacturer's mark on the back of a cockade.

◀▲ Reverse of the national emblem. It is a model adopted in 1935 made of aluminum or zinc. The initials denote the manufacturer, "38" indicates the year of manufacture.

▲ Different insignia on Schirmmützen adopted in 1935; the national emblem and the cockade of the Reich combined with a wreath of oak leaves. Gold-colored insignia appeared in 1943 and was intended for Heer generals' caps.

OFFICERS

HEER OFFICERS:

When on parade or campaign, officers were allowed to wear the Schirmmütze. Officers had to purchase their own caps and uniforms. From the start of hostilities and for economical reasons, officers were only allowed to buy a maximum of two caps.

Officers generally had a preference for peaked caps on offer, such as the police cap, and wore them all the time; except when a helmet was necessary.

▶ **This smiling Hauptmann, probably infantry, offers an excellent view of his Alter Art Feldmütze with its embroidered insignia. This officer's cap has a leather visor and was adopted in 1934.** (Collection M. Zauner)

▲ **Alter Art "combat" cap for an infantry officer.** This cap is made of field-grey material; on its dark green band the eagle and oak-leaf wreath are BeVo woven. The metal Reich cockade was usually pinned on. The visor is supple leather, and the lining is green.

▶ **The two officers in the middle are wearing the Alter Art Feldmütze. The two NCOs on either side of them are wearing traditional Schirmmütze and modified Czech tunics.** (Collection Hofmann/Lötsch)

▲ **Oberarzt Dr. Franz Peter Rapf of the 9th Panzer Division. The doctor's Schirmmütze closely resembles the Alter Art Feldmütze by its shape and insignia, except that a silver officers' braid has been added, transforming it to a peaked cap. As seen, Rapf was highly decorated: Iron Crosses 1st and 2nd Class, 1st Winter on the Eastern Front Medal, Wound Badge in Silver, Infantry and Panzer Assault Badges, Hitler Youth Badge.** (Collection Ludwig Bauer).

▲ **Alter Art Artillery Officer's cap,** made by Albert Kraus. The visor is made of varnished leather; the insignia is woven aluminum thread reserved for officers. The interior is orange canvas. (Private Collection)

▲ **Alter Art Schirmmütze for Pionier Officers** in field-grey serge, dark green band, black velvet piping, silver embroidered insignia stitched through the inner lining. The latter is made of silk with an interior celluloid sweatshield marked, "Wilhelm Welhausen Kassel." The sweatband is made of leather. Size 59½.

▲ Variants of the national emblem, woven or embroidered in silvered wire.

▲ **Schirmmütze of a Veterinary Officer.** Field-grey serge, silver embroidered insignia, carmine piping. Lining gold colored. Size 55.

▲ Leutnant Doppelhofer on the rifle range with his Mauser 98K. He is wearing an "Alter Art" officer's M1934 Feldmütze which, after 1938, was no long available for purchase by younger officers.
(Collection Hofmann/Lötsch)

OFFICERS

▶ Watched by his amused men, Oberleutnant Florian tries out a Mauser 98K. He wears a saddleback Schirmmütze with metal insignia. (Collection Hofmann/Lötsch)

◀ Artillery Officer's cap with red piping. Yellow cotton lining, oak-leaf wreath embroidered in silver wire, and aluminum eagle. Size 57.

▲ Cockade variants with the oak-leaf wreaths either woven, embroidered or metal. The first is an early metal fabrication. The last is embroidered and fabricated at the end of the war.

▲ **Schirmmütze of an Artillery Officer** in field-grey serge with red piping and aluminum insignia. The lining is yellow silk and is stamped "Zimpel Einem."

▲ **Officer and general staff chinstraps.**

▲ **Heer Medical Officer cap.** Field-grey serge, dark green band, dark blue piping, metal eagle, embroidered oak-leaf wreath, silver braid. The lining is white, and on the celluloid sweatshield is the nazi party eagle.

▲ This infantry Hauptmann, who has all the characteristics of a veteran of the front, has kept his cap with its metal insignia intact: without damaging it or modifying it in any way. He wears the German Cross in Gold, Infantry Assault Badge, Wound Badge in Black, plus a ribbon bar including the Iron Cross 2nd Class, Medal for the 1st Winter on the Eastern Front and two army service ribbons.

▶ **Heer Panzerwaffe:** shown here are two close-fitting, double-breasted tunics (also known as the panzer wrap) with pink piping, and the distinctive collar tabs of the panzer forces – rectangular on a black background with pink piping and the Death's Head (inherited from the Prussian Hussars) – a leather belt with tongue and large rectangular buckle. Our Oberleutnant wears a Schirmmütze for panzer officers in field-grey serge with pink piping with insignia embroidered in silver wire; he is carrying long binoculars reserved for the Panzer units and artillerymen. The figure at right shows a member of the troops with the black Panzer wrap, but this model is after 1940 (no pink piping around the collar), though the pink is still present around the collar tabs and the shoulderboards. He has a cap of black serge with pink piping and with the insignia woven on a black background. Completing the ensemble in front are a standard issue cap in field-grey serge (orange canvas lining) for NCOs and troops, earphones and throat microphone in the original box, a chart identifying Russian and Anglo-American tanks, a wooden teaching model of a Tiger I tank, a pair of shoulderboards, and a collar tab characteristic of the panzer forces.

OFFICERS

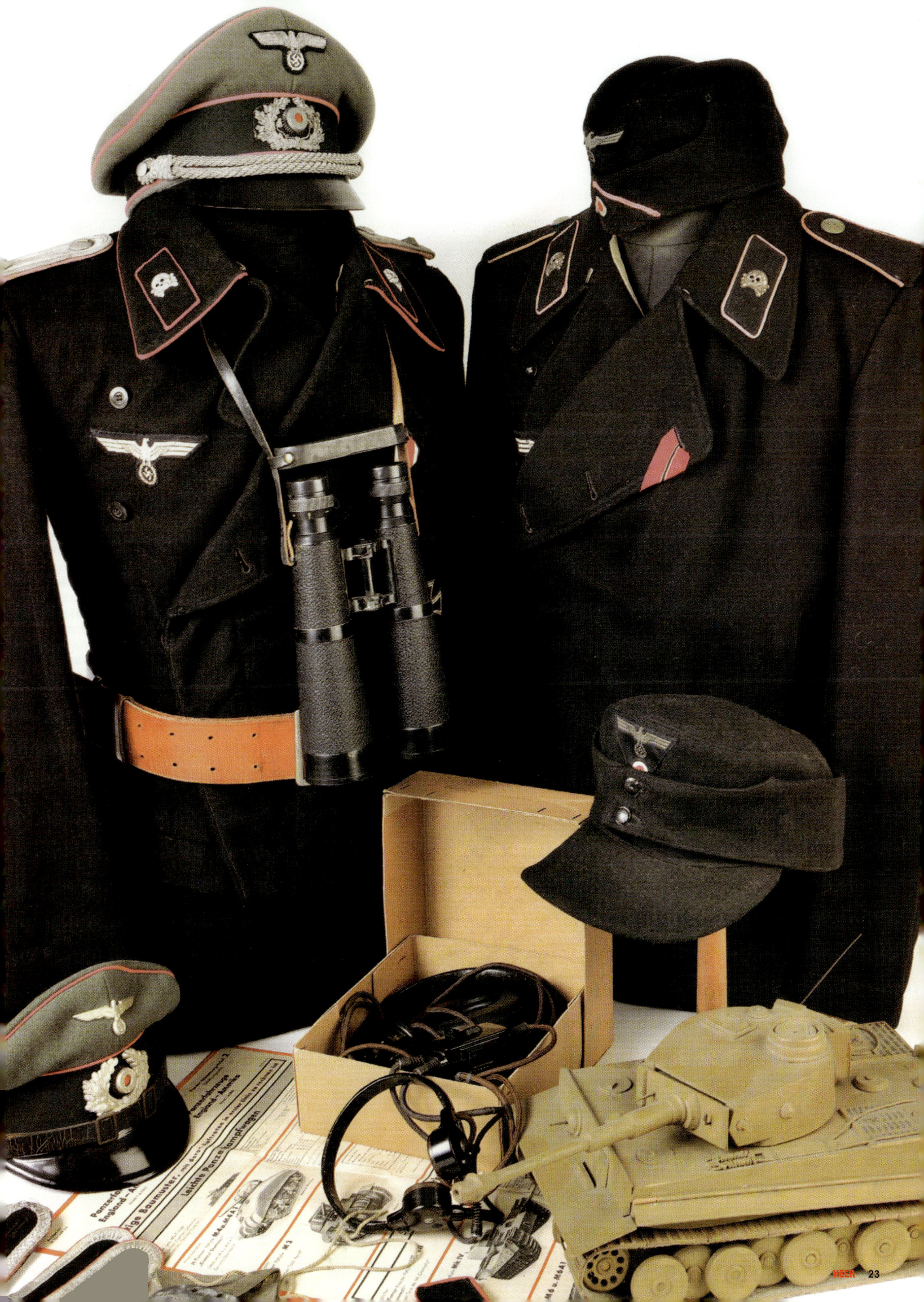

▲ This highly decorated Hauptmann wears an officer's Schirmmütze with metal insignia. Among his decorations are the German Cross in Gold, the Iron Cross 1st Class (curiously he has no ribbon for the Iron Cross 2nd Class in his buttonhole), the Infantry Assault Badge and the Wound Badge in Black. As for his friend at right, he is wearing an Assault Badge, Wound Badge, and the prestigious Knight's Cross. Note on the man at left the enlarged collar with NCO's stripe, and the embroidered eagle above his right breast pocket.

▲ **Cap for Staff Officers** made of field-grey serge with carmine piping and metal insignia. The lining is yellow silk and marked, "EREL Sonderklasse Extra."

▶ **Cap for an Officer of the Nachrichtentruppe**, identifiable by the lemon yellow piping. The eagle is made of aluminum and the oak-leaf wreath is embroidered in silver thread. The manufacturer is "Wilhelm Wellhauser, Hannover."

OFFICERS

▲ This infantry Hauptmann is a veteran of World War I and a member of the Nazi Party. He wears the Iron Cross 2nd Class ribbon in his second buttonhole, and the Iron Cross 1st Class on his breast pocket. Among other decorations are the 1929 Nuremburg Rally badge, commemorating the "4th Year of the Reich Party," which had been worn as an honorary insignia before becoming official in 1936. He also wears the NSDAP Long Service Award, and an SA Sports Badge. His high-peaked Schirmmütze displays the wreath of oak-leaves and the eagle both embroidered in silver thread. Note that the wings on the national emblem are unusually long. (Collection XA)

▲ **Schirmmütze for Officers of the Feldgendarmerie.** The insignia is in metal and the piping is orange. The lining is golden silk marked with, "Aug. Braun – Marburg." (Collection FC)

◄ **Schirmmütze for an Officer of the Kraftfahrtruppe,** in field-grey wool with a dark green band, light blue piping and metal insignia. The lining is yellow silk, and the celluloid sweatshield is stamped, "Die Qualitatsmütze – Joh. Seitz Landau."

▲ Cap for an Officer of Signals: field-grey serge with yellow piping. The lining is white silk, and is manufacturer marked, "Klementz/Insterburg." The insignia is in metal. Size 57.

▲ Leutnant Jungwirth wears his newly awarded Panzer Assault Badge, pinned for the first time on his tunic. His cap has pink piping, and insignia embroidered in silvered wire. (Collection Hofmann/Lötsch)

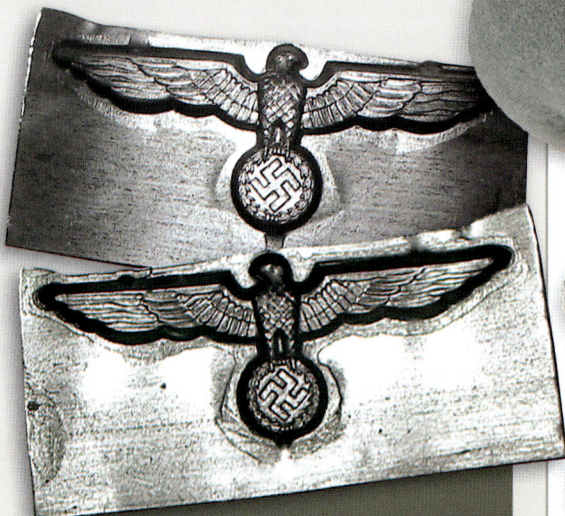

▲ Mold for making the national emblem.

▲ Cap for an Officer of the Gebirgsjäger. This cap is made of field-grey serge with green piping and aluminum insignia; the Edelweiss is attached between the eagle and the oak-leaf wreath. The lining is yellow silk with the maker's marking, "EREL Sonderklasse Extra." Size 56½.

◄ **Schirmmütze for an Officer of Pionier;** field-grey with a dark green band, black piping, metal insignia and silver braid. The lining is orange and on the celluloid sweatshield is marked, "Pekuro Stirndruckfrei." The cap has a leather sweatband.

◄ A superb **Cavalry Officer's Schirmmütze,** in mint condition; it has crossed the years without being damaged in any way. It is made of fine field-grey serge with a dark green band, aluminum insignia, silver braid and golden piping. The lining is golden silk. On the leather edge is stamped, "EREL Stirnschütz D.R.G.M., D.R.P. Angem." Stamped on the celluloid sweatshield is, "Offizier Kleiderkasse, Berlin, Ges Gesch EREL Berlin, Sonderklasse Extra." The small piece of paper that goes with the sweatband's label states: "EREL Stirnschütz DRGM bleibt Stirndruckfrei. Deise mütze bietet auberdem: Wirksame ventilation durch spezial kokarde oder Abzeichen, Durchschwitzen des Mützebundes wird vehindert." Translated, this bit of advertising reads: "EREL protection that doesn't press against the forehead. Among other things, this cap offers: efficient ventilation through a special cockade or insignia that prevents the band becoming soaked in sweat." On this cap the patented ventilation grid is clearly visible in the center of the Reich cockade. This cap belonged to Major Freiherr von G. of the 5th Brunswick Cavalry Regiment (the same regiment as Claus Schenk Graf von Stauffenberg). The travelling box is the original and was called, "die Mützenschachtel" (cap box). There are a few variants; some rectangular such as "der Mützenkoffer" (cap case). However, they were considered a luxury item that many could not afford.

(Collection XA)

▲ This young officer wears a superb Schirmmütze with an eagle whose wings are much longer than other embroidered models.
(Collection Franco Mesturini-Luc Soldati)

▲ **Infantry Officer's Schirmmütze,** in fine wool with a dark green band, white piping and metal insignia. The lining is gold cloth. Size 54.

GEORG RIEDEL
Anfertigung sämtlicher
STICKEREIEN
FÜR
HEER U. MARINE
GEGR.1879
NO 55 GREIFSWALDERSTR. 224
TEL. 533405
SPEZ.
...OMATEN-
...EREIEN
...ÄNDER

▲Georg Riedel. Complete embroidered creations for the Army and Navy, founded in 1879. Specialist in embroidery for all foreign diplomats." Note the represented figures from left to right: a diplomat, a Kriegsmarine officer, and a Heer general.

▲ **Truppensonderdienste Officer's cap.** This cap is made of field-grey serge with a dark green band, dragon-blue piping (appearing in 1944), silver embroidered oak-leaf wreath, metal eagle and silver braid. The lining is golden with an ersatz material sweatband.

▲ **Panzergrenadier Officer's Schirmmütze** made of field-grey gabardine with a dark green band, grass-green piping and metal insignia. The ersatz visor material has been given a black gloss. The lining is a golden material with the stamp, "Offizierskleiderkasse Berlin – EREL Sonderklasse Extra." Size 59.

▲ Young infantry Leutnant decorated with the Iron Cross 2nd Class. His Schirmmütze, with its white piping, has an oak-leaf wreath embroidered with silvered wire, as well as metal insignia.

▲ **Propagandakompanie Officer's Schirmmütze** made of field-grey wool with a dark green band, light grey piping, metal insignia and silver braid. The lining is silver grey with the silver stamp "Drescher Schweinfurt." The sweatband is also grey.

▲ Junior officer wearing a cap with Pionier black piping. (Collection Franco Mesturini-Luca Soldati)

▲ **Cavalry Officer's Schirmmütze.** The insignia is embroidered in silvered wire, and the piping is golden. (Private collection)

▶ Senior officers of the 15th Artillerie-Regiment in greatcoats and Alter Art officer's caps.

▲ **Nebelwerfer Officer's Schirmmütze.** Made of field-grey wool with burgundy piping and aluminum insignia. The lining is black silk with the stamp, "Roland Mütze." The sweatband is made of ersatz material. Size 55.

▶ On the left, a **Panzergrenadier Hauptmann,** wearing a summer issue tunic, made of lightweight canvas. The shoulderboards and collar tabs are piped in panzergrenadier grass-green, the ribbon is the award for the 1st winter spent on the Eastern Front, and the Iron Cross 1st Class. The cavalry-twill Schirmmütze with stamped metal insignia (aluminum in this case) is also piped in grass green. On the right: another Hauptmann, but of the 1st Infantry Division of the Heer, white piping, with a M1933 tunic, modified in 1936 (collar dark bottle-green) and the insignia of an officer. The Infantry Assault Badge in Silver is pinned to the breast pocket. Also included are a troop belt (as demanded in the campaign regulations) and a blackened leather holster for the P.08. His Schirmmütze is tailor-made with its insignia in silvered metal; our officer has removed the silver braid to give his cap the look of an Alter Art campaign cap; he has pinned his division's insignia on the left side of the band (on vehicles the shield is topped by a crenellated bar). On the stand in front is a piped cap for an infantry officer, and at right is another pink piped for Panzer or Panzerjäger. Also included are a vehicle pennant, a cavalry or reconnaissance officer's cap. A small, wooden halftrack has another Schirmmütze resting on it. With its purple piping this cap would normally belong to a military chaplain, but as it does not have a Latin cross, it could be a deacon's cap. In front are shoulderboard examples with various waffenfarbe. (Private collection. Photo Frederic Coune)

▲ **Detail of the 1st Infantry Division badge.** (Private collection)

▲ **Pionier Officer's Schirmmütze.** Made of field-grey gabardine with a green band and black piping, metal insignia and silver braid. The lining is green silk with the following stamped in silver letters, "Offizier Kleiderkasse Berlin – EREL Sonderklasse, Extra."

▲ Taken by a professional photographer, just after being promoted to Leutnant and Zugführer of 1./Panzer Regiment 33, Ludwig Bauer poses in a uniform that strictly complies to Heer regulations. The insignia on his brand new Schirmmütze are totally embroidered in silvered aluminum wire: oak-leaf wreath, Reich cockade and national emblem. The ribbons in his second buttonhole represent the awarding of the Iron Cross 2nd Class and the Winter Campaign on the Eastern Front. Next to the Panzer Assault Badge, the Wound Badge is a version awarded to soldiers who fought in the Spanish Civil War (1936-1939). The snaphook, and its leather support, seen on his left hip, is for attaching the officer's sword. Officers were also permitted to carry a bayonet or dagger as sidearms. However, on December 23, 1944 these arms were forbidden throughout the Wehrmacht and were replaced by an automatic pistol. (Collection Ludwig Bauer)

▲ **Cavalry Officer's Schirmmütze** made of field-grey wool; the metal insignia here are separate. The lining is green silk; the sweatband is made of ersatz material.

▲ **Artillery Officer's Schirmmütze.** This cap is olive green and intended for operations in the Mediterranean theater. It has a dark green band and red piping; the insignia is metal and it has a silver braid. The lining is blackish-grey cotton and the sweatband is grey.

▲ An Oberleutnant in his summer uniform, adapted for operations in the Mediterranean theater. Like his tunic, his Schirmmütze also appears to be made of a lighter material than usual; it has a metal eagle and cockade, the latter surrounded by the embroidered oak-leaf wreath. This Oberleutnant has been decorated with the Iron Cross 2nd Class and the Wound Badge in Black.

▲ 116th Panzer Division cap insignia.

▲ **A superb Panzerwaffe Officer's cap,** that once belonged to Heinrich Tempeler, Oberleutnant der Reserve, commander of the 3rd Company of the 21st Panzer-Abteilung. He was killed on July 9, 1943 in Samodurowka, Russia. The cap is wool with the insignia entirely embroidered in silvered wire, and is piped in panzer pink. On the celluloid sweatshield is the following inscription, "Offizier Kleiderkasse Berlin, EREL Sonderklasse, Extra"; this cap therefore came from the officer's clothing store. "Extra" means superior quality from EREL and "Sonderklasse" means special class. The lining is light green silk, an unusual color for this type of cap.

▲ **A superb quality cap for an Officer of Infantry Regiment 17.** The oak-leaf wreath is embroidered silvered wire, the cockade is metal. The national emblem and the Brunswick Death's Head are silver metal. The piping is infantry white. (Collection XA)

▲ **Infantry Officer's Schirmmütze.** The manufacturer's stamp is on the yellowish lining, "J. Lettel, Nikolaistrasse 18B Hannover." It has a diamond celluloid sweatshield. (Collection XA)

▲ This junior officer, photographed in 1936, is wearing a magnificent Schirmmütze with an eagle whose wings are longer that the later model. (Collection Franco Mesturini-Luca Soldati)

◀ **Panzergrenadier Officer's cap.** The piping is grass green and the insignia is stamped aluminum. The inner celluloid sweatshield has the profile of a soldier wearing a cap and the inscription, "Deutsche Wertarbeit" ("Quality German Workmanship"). (Collection XA)

▲ **Two brothers pose before leaving for the front.** This Leutnant of the Panzergrenadiere already has the Iron Cross 2nd Class and the Wound Badge in Black. It has been meticulously rubbed to look like a battle-worn insignia and perhaps even a badge of a higher class. His Schirmmütze has been modified to an Alter Art (with corresponding woven insignia) which he probably had the habit of keeping in his pocket. His brother at left has a Schirmmütze cockade on his M43 Feldmütze. (Collection Frieder Ortlepp)

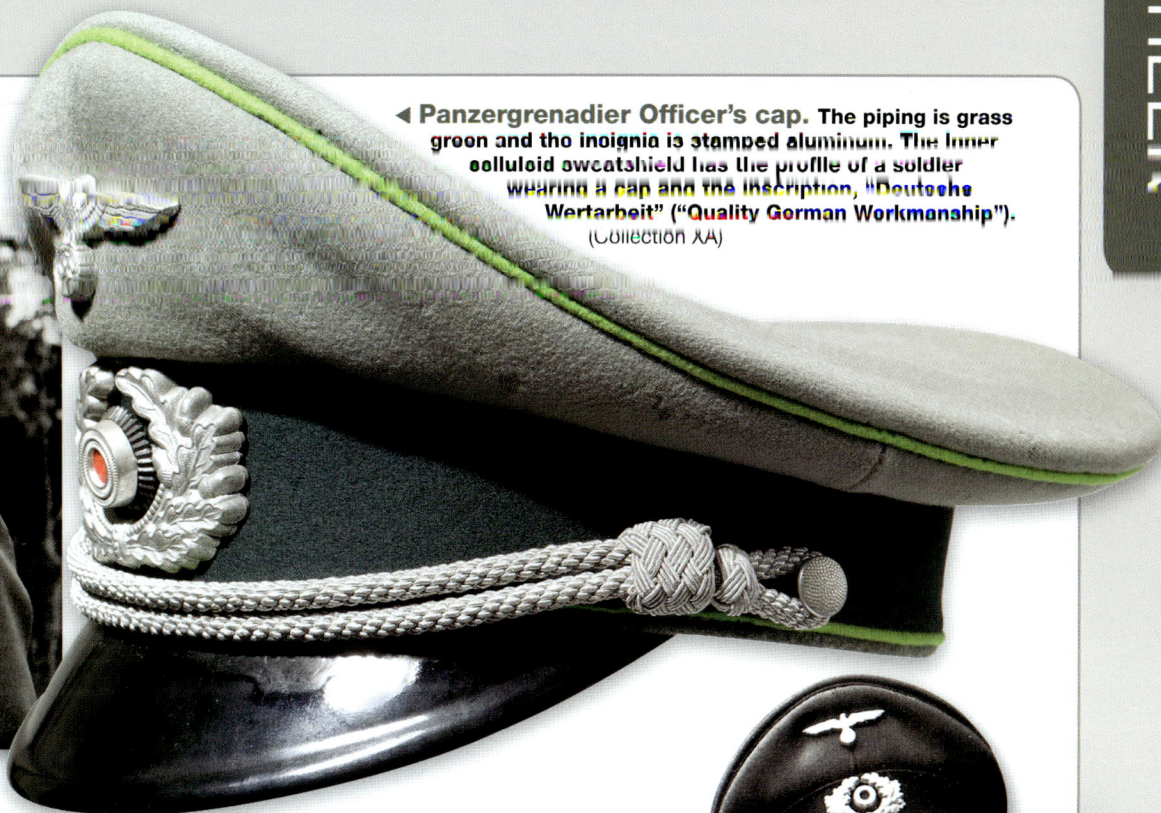

▶ **Catholic Chaplain** wearing a classic Schirmmütze with the purple piping of the clergy. Catholic chaplains differed from their Protestant brothers by the figure of Christ on the cross around the neck, while Protestants wore a simple cross. (Collection Franco Mesturini-Luca Soldati)

▶ **Military Chaplain Officer's cap** (either Catholic or Protestant): pastor and senior pastor. The Feldbischöfe (there were only two "Field Bishops") were general officers. The chaplains finally received a definitive uniform in 1937: field-grey with matching cap, which had purple piping, a dark green band, a Gothic Cross between the oak-leaf wreath and the eagle. In this photo the cross is embroidered in silvered wire, but a metal model also exists that can be pinned to the cap. The Feldbischöfe's cap had golden piping (like Heer generals); except for the piping above the band, which was purple. (Collection FC)

▶ **Generalmajor Horst Kadgien's cap**

Horst Kadgien (1897-1980) commanded the 36th Infantry Division and was wounded on January 18, 1944; On March 1, 1945 he took over the Heer's Artillery II School. This photo is of a high quality EREL cap, with its braid woven in gold wire, surprisingly fancy flat chinstrap buttons and its gold eagle (smaller by a third than the standard model, but not of the first model). This cap with its particular characteristics is easily recognizable in several photos form the era. In addition, this cap is made of fine field-grey serge with a dark green band, its piping and insignia are gold, as befits a general officer. The lining is green with a silver inscription on the celluloid sweatshield, "EREL Sonderklasse, Extra" (EREL Ges.Gesch. Berlin, Sonderklasse, Privat). The inside edge is brown leather with letters stamped in gold: "EREL Patent Stirnschultz." In a slot in the celluloid sweatshield is a small calling card with the owner's name: "Generalmajor a.D. Kadgien." If this is undeniably the original, one should be careful: manipulation being easy, it is possible to increase the value of the object due to the presence of a found or copied calling card.

◀ **Generalrichter Schirmmütze** in fine wool with a dark green band; the piping is golden, except above the band, where it is the wine red judicial color. The insignia are silvered, the braid is gold. The lining is brown silk with a leather sweatband. This model cap is very rare. Size 57.

▲ **Beamte Diensmütze** from the Reichswehr period. This standard issue cap was privately purchased. It is made of field-grey serge with dark green band and piping. The insignia is metal: cockade, early model of the oak-leaf wreath and the eagle. The lining is brown with a silver inscription, "Stirndruckfrei – Deutsches Reichspatent" ("Does not press on the forehead, German patented invention."). Size 57.

▲ **Heeresverwaltung Clerk's** cap (officer rank): made of field-grey serge with dark green band and piping, metal insignia and silver braid. The lining is either grey or gold with the celluloid sweatshield stamped, "Deutsche Wertarbeit." The sweatband is made of ersatz material. The Wehrmachtbeamte were officials such as paymasters administrative staff, etc. Their generals' caps had gold piping, like regular Heer generals, but above the band the piping was green. Also, Heer officials "auf Kriegdauer" (for the duration of the war – cancelled in 1943) had blue-grey bands instead of dark green. They were civil servants in uniform, working in territories occupied by the Reich.

▲ This general, photographed before the war, wears a Schirmmütze with the short-winged 1st pattern eagle. (Collection Franco Mesturini-Luca Soldati)

▲ This Heer official is wearing a Schirmmütze specific to his rank as Administrative Official. (Collection Franco Mesturini-Luca Soldati).

▲ **Heer Official's cap** (officer rank). Made of field-grey felt with dark green piping. The insignia is stamped aluminum. Made by, "Dalüge, Hannover."

▲ Administration Official (Major). The collar tabs and shoulderboards are specific to this branch. This photo was taken before the war. He is wearing a shoulder belt that was no longer worn after the Polish campaign in 1939 (as it allowed the enemy to easily spot an officer). He also wears a nice private purchase cap with the cockade and oak-leaf wreath embroidered in aluminum wire, a metal eagle and dark green piping. (Collection FC)

▲ **Sonderführer's Schirmmütze.** Made of field-grey wool with grey band and piping and metal insignia. The lining is yellow silk with silver inscription, "EREL Sonderklasse Extra, Offizierskleiderkasse Berlin." Size 57½.

OFFICIALS

LUFTWAFFE

The Luftwaffe adopted the Schirmmütze and added its own touches, such as the winged Reich cockade and cap badge. The highly prestigious fighter pilots with their aerial victories allowed themselves several fancy items to highlight their status as combat veterans – these flyers were in everyone's imaginations and they became legendary. One might also ask oneself if the Luftwaffe's commander, Hermann Göring, wasn't also the first to promote the archetypical Luftwaffe officer's look, with his own formidable, and sometimes improbable, embroidered caps.

NON-COMMISSIONED OFFICERS AND TROOPS

▲ **Schirmmütze for Mannschaften/Unteroffiziere der Fliegertruppe (Navigators and Paratroops).** The insignia is made of aluminum, the band is black Mohair, and the piping is yellow. The lining is also yellow. Size 56.

▲ **Oberfeldwebel or Stabsfeldwebel.** The eagle on his cap is an early model.

◄ **Luftnachrichtentruppe cap.** This cap is blue-grey with a black Mohair band and light brown piping. The eagle and oak-leaf wreath are stamped metal. The lining is yellow silk with the inscription, "Deutsche Wertarbeit Stirndruckfrei" ("Quality German Workmanship"). Size 56½.

▶ **Veterinarian Mannschaften/ Unteroffiziere Schirmmütze.** Blue-grey serge with black Mohair band and carmine piping. The lining is blue-grey cotton and stamped, "Ideal." The insignia is aluminum and the chinstrap is glossed leather. Size 56.

▶ **An NCO, veteran of the First World War in conversation with an officer. Each wears a cap corresponding to his rank.**

▲ **This young Unteroffizier is already a veteran as shown by the Iron Cross 2nd Class ribbon in his buttonhole, and the ribbons representing the Westwall Medal and the 1st Winter in Russia. He has not modified his privately purchased high peaked cap. His shirt does not strictly comply with regulation wear.**

▶ **NCO and Fliegertruppe cap,** summer wear. The white crown cover is removable, the band is black mohair, the insignia is aluminum, and the piping is yellow. The lining is white silk, stamped, "EREL-Strinschutz." Size 56½.

▶ Rare Luftwaffe **Technical NCO and Troop Schirmmütze.** Privately purchased woolen model with a black mohair band, pink piping and aluminum insignia. The lining is blue cotton. Size 59.

▶ This NCO is a veteran of the Crete Campaign and the Eastern Front. As a navigator his cap has yellow piping.

▶ **Luftwaffe Sanitätstruppe Mannschaften/ Unteroffiziere Schirmmütze.** The blue piping belongs to the Health Service, the insignia is made of aluminum. The lining is gold colored. Size 56.

▶ **"Hermann Göring" Division NCO and Troop cap.** Blue-grey material with black mohair band and white piping. Leather sweatband and yellow lining, both of which are stamped, "Firma EREL." The insignia is made of zinc. The visor did not receive a properly finished edge. Size 56.

▲ **White Schirmmütze** with light green piping denoting Luftwaffe radar units.

▶ At left is a **Luftwaffe Paratrooper Leutnant:** the same corps colors – golden yellow – as navigators on the collar tabs and shoulderboards of his tunic, which is made of gabardine (made by a master tailor). Our officer is the holder of: the German Cross in Gold (here in its embroidered version, stitched to the), Paratrooper Badge, and the Iron Cross 1st Class, as well as the prestigious Knight's Cross. He wears a nice example of the Luftwaffe officer's Schirmmütze and he has somewhat changed the shape. The insignia is embroidered with silvered aluminum wire. This gives the cockade a pronounced effect; the edge of the visor is made of mica. To finish off our paratrooper's gear is a late-1939 model paratrooper helmet: it does not have the national emblem, the screws are aluminum, and it has no ventilation. The regulation netting is the original. Next to it is a Luftwaffe Fliegermütze also made by a master tailor in gabardine and silvered wire embroidered insignia. There is also a Luftwaffe officer's dagger. The figure on the right is a fighter pilot NCO, wearing a superb navigator piped NCO Schirmmütze with early aluminum insignia. He is wearing an NCO flying tunic over which he has added a warm leather jacket, nicknamed "Defense of the Reich." There is also a pair of blue-grey suede flying gauntlets, a Flugzeugführerabzeichen (pilot badge) in its original box, and a pilot's sleeveband denoting fighter unit Jagdgeschwader 3 "Udet." (Private collection. Photo Frederic Coune)

▲ **Reichsluftwaffenministerium NCO and Troop cap.** This Schirmmütze is made of gabardine and has black piping and aluminum insignia. The lining is brownish-yellow rayon with an ersatz material sweatband.

▲ **Two Oberleutnante deep in discussion.** The officer at left wears a high fronted "Teller form" cap. At center appears to be Günther Korten who was the head of the Luftwaffe general staff during the Polish Campaign in 1939, and who was later promoted to Generaloberst posthumously (he died from wounds received during the July 20, 1944 assassination attempt on Hitler).

▲ **Landespolizeigruppe "General Göring" NCO and Troop Schirmmütze.** Polizei green material with a dark green band, light green piping and aluminum insignia. The lining is yellow silk with, "Friedrich Sackmann Berlin" stamped in gold. This is a very rare cap. Size 58.

▲ **1940:** Adolf Galland while he was still a Major and commanded Jagdgeschwader 26 "Schlageter." He is wearing a splendid "saddle-back" Schirmmütze, tipped slightly to the left. He has received the Spanish Cross with Diamonds, the Iron Cross 1st Class, and the Knight's Cross. The Leutnant behind him wears a Fliegermütze with silver piping denoting an officer. (DR)

▲ **Luftwaffe Officer's Schirmmütze.** Grey-blue serge with a black mohair band and the officer's silver piping (a Reserve Cross, "Reservekreuz", was replaced by a cockade), and silver braid. The lining is yellow silk with a mica manufacturer's stamp, "Verkaufsabteilung der Luftwaffe Berlin – Erstklassig" ("Commercial Sales and Service to the Luftwaffe"). The sweatband is leather. Size 54.

▲ The insignia on this Leutnant's cap is embroidered with silvered wire. The eagle is an early model with short wings and raised tail-feathers.

▲ **Jagdgeschwader 2 "Richthofen" pilot in Triqueville, France, 1942.** Major Erich Leie poses in a way that shows his pilot's uniform to its best advantage: Knight's Cross around his neck, "cyclist's" jacket, and life vest which has seen a lot of use. The piping and insignia on his Schirmmütze are in silver.

▲ Luftwaffe fighter ace Oberleutnant Adolf Dickfeld wearing the Knight's Cross with Oak Leaves.

▲ **Luftwaffe Officer's cap.** The insignia is embroidered with silvered wire; the braid and piping are also silver.

▶ Leutnant Otto Gemünden in a studio portrait, May 1943. His cap has kept its original "strict" shape. The insignia is in the classic officers' silver-wired embroidery. He has also received the Knight's Cross, the Wound Badge (in Gold or Silver), and the Flak Badge (DCA).

▲ **Medical Officer-candidate's cap.** This summer cap is made of lightweight gabardine. A Luftwaffe officer-candidate counts as an officer and has the right to wear insignia embroidered in aluminum and a silver braid on his hat, but he keeps the blue piping of the medical corps.

▲ **Luftwaffe Officer's Schirmmütze in white.** The crown cover is removable; the piping is silver denoting officer rank. The national emblem, cockade, and oak-leaf wreath are all embroidered.

▲ Piping and insignia detail on a Luftwaffe Schirmmütze.

▶ During a briefing: The officer on the left, a Luftwaffe Leutnant (horizontal bar surmounted by wings on his sleeve – an insignia for flying suits that didn't have shoulder boards) wears a white cap reserved for the summer uniform in peacetime but also worn during the war. This cap for junior and senior officers has silvered-wire embroidered insignia.

▲ **Knight's Cross recipient Leutnant Erich Jeckstat's summer cap.** The white crown cover is removable, the band is black mohair, the piping is silver, the Luftwaffe flying eagle insignia is metal, the oak-leaf wreath and cockade are embroidered with silvered wire, the braid is silver. The sweatband is made of leather.

OFFICERS

▶ On the right: the figure of a **Luftwaffe Officer Pilot** wearing a tropical troop tunic with officer's shoulderboards, and as can be seen in a lot of photos of the era, officers added their rank to their collar, a ribbon representing the Iron Cross 2nd Class in the buttonhole and Pilot's Badge on the left breast pocket. Our officer wears a summer Luftwaffe officer's Schirmmütze with removable white crown cover.

◀ On the left: **An officer of the Hermann Göring Division in Tunisia;** this is the only division that was equipped with camouflage smocks as well as the camouflage cap cover used by the Waffen-SS. Under his camouflage smock our officer is wearing a classical tropical shirt; he is also wearing a rare Luftwaffe tropical cap (called a "Ramcke"). This is an original model, with its braid of woven aluminum thread and undoubtedly worn by an officer; a lot of period photos confirm this type of cap. This particular cap was brought back by a French veteran of the Tunisian Campaign in 1943. Below: a wooden sign in the shape of a command pennant in the color of a flak unit and very likely bearing the name of its commanding officer, "Walbrach." M1940 Luftwaffe helmet with spray painted, sand yellow camouflage and standard netting, Luftwaffe enlisted forage cap in its tropical version made of sand colored canvas; a wooden plaque depicting a Luftwaffe soldier in front of his sentry box; a Medical Officer-candidate's Schirmmütze (who has officer's rank and therefore the insignia and braid of woven aluminum wire, but with the blue piping of the medical corps). Collar tabs and shoulderboard of a medical officer, and a metal eagle for an officer's summer or tropical uniform round out the ensemble.

◀ **Luftwaffe Officer's Schirmmütze.** The white crown cover for summer use is removable (for easy cleaning). It came from the Luftwaffe Sales Service, Berlin SW 68, Puttkamerstrasse 16-18, Erstklassig. Size 54½. (Private collection)

▲ **General Officer's Schirmmütze.** This grey-blue cap belonged to Generalmajor der Flak-Artillerie Adolf Wolf – he defeated General de Gaulle at the Battle of Abbeville and received the Knight's Cross (The cap was sold by the Generalmajor's family). It is a fine example of a saddle-form cap; the insignia, piping and braid are gold wire. It was made by EREL for the officer's uniform store and bears the quality "Extra" mark (which usually means that the red cockade in the center featured the EREL patented ventilation grid). Wolf's signature and his rank are marked in ink on the inside edging. (Collection XA)

◄ The cap shown here belonged to Reichsmarschall Hermann Göring, who was the only person to wear a cap in this light grey color. The embroidery is also unique and of excellent quality. The laurel leaves are embroidered in gold and the band is grey velvet; the swastika is partially unstitched. This prestigious cap was delivered by EREL in August 1940.

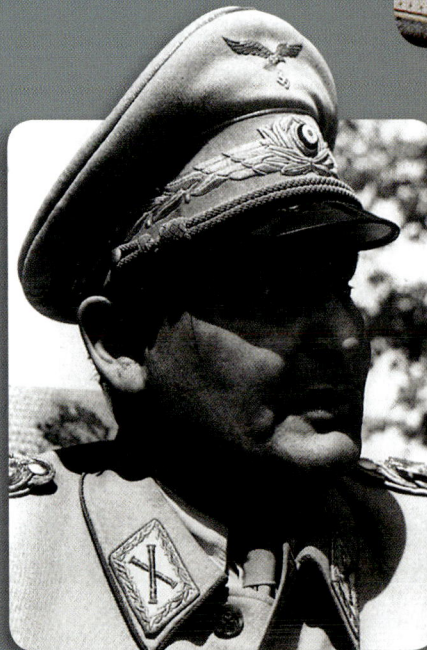

◄ This photo shows off the laurel leaf gold embroidery and the grey band.

REICHSMARSCHALL HERMANN GÖRING

▶**Luftwaffe General Officer's cap.**
This Schirmmütze has the white summer covering. It belonged to Ernst-August Roth, who served in the navy during the First World War and who, later, in the Luftwaffe, commanded at Bordeaux, Mérianne, Nantes, Beauvais, Sicily, Denmark/Norway during the Second World War. He was the last Luftwaffe commander in Norway in 1944/45 and received the Knight's Cross. This cap was discovered a few years ago in Charente-Maritime, in a garage in Tremblade. The embroidered cockade does not have the ventilation grid usually found in "Extra" quality caps made by EREL. Made of white cotton canvas with a slightly "waffled" appearance, this model was adopted in 1937. In a slot in the diamond-shaped sweatshield is a small note with the name of the cap's owner, probably in his own handwriting. Luftwaffe Clothing Store, Berlin SW68 Putkamerstrasse 16/18, EREL Sonderklasse, **Privat.** (Collection XA)

▲ Knight's Cross recipient Generaloberst Alexander Löhr, shown here on August 4, 1941 while he was commander-in-chief of the Balkans.

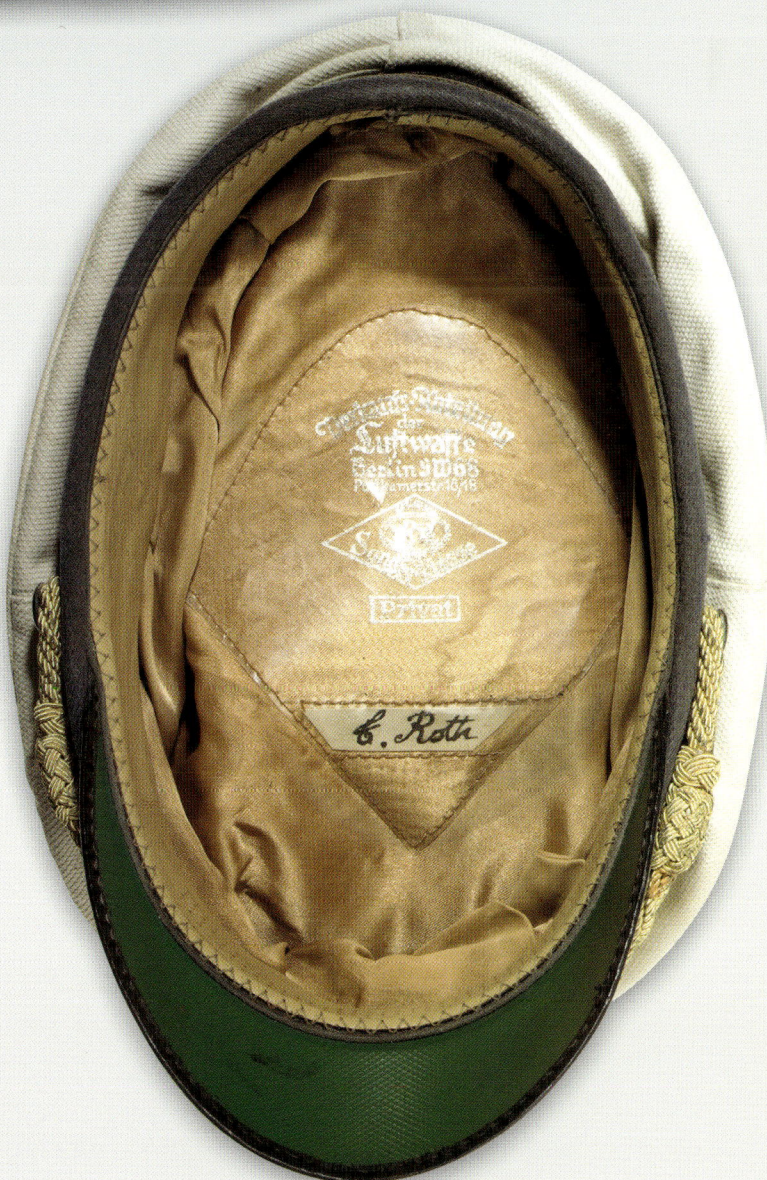

▲ General von Schroeder of the Flak-Artillerie wears a magnificent cap with gold insignia.

▲ **Luftwaffe General's Schirmmütze** in blue-grey serge with black mohair band; the insignia, braid and piping are embroidered celleon (nylon). The visor is glossy black without a border, the lining is light green silk, and the sweatband is leather. Size 57½.

▲ Generaloberst Kurt Student, commander of the Fallschirmjäger, wears a cap with gold wire insignia.

KRIEGSMARINE

The Kriegsmarine was the Third Reich's naval force from 1935-1945. The year 1935 was marked by the adoption of the new warship with the swastika common to all the German forces known under the name of Wehrmacht. The German navy uniform, even if it resembled that of many other naval forces around the world with its gold and blue, had certain characteristics of its own: between those inherited from the earlier Imperial Navy and those, more modern, installed by the now commanding powers. As we will see, the Schirmmütze was a part of these changes, and with its form and the modifications made by those who wore them, it holds an important place among the mythic caps of the Third Reich.

The Schirmmütze was identical for all officers and petty officers. The visor was leather, the band black mohair and the crown cover either blue or white (for the summer) but more supple than Heer caps. The chinstrap was affixed by two small buttons, with an anchor motif at the center. The eagle and the swastika were embroidered in yellow cotton or gold wire (replaced by a gold metal eagle for the summer cap) on the upper part of the hat. The national red, black and silver cockade, surrounded by the oak-leaf wreath, was affixed to the band. The visor was black leather without trimmings for petty officers; junior officers had blue material visors with a thin gold band around the edge representing waves; senior officers from Korvettenkapitän to Kommodore wore a band of golden oak leaves 18mm wide, and Admirals wore two wide bands of golden oak leaves. U-boat commanders were allowed to wear the white crown cover (either fixed or removable), which was usually reserved for summer or for navigating in tropical waters. A number of these caps were made by private hatters, so the material used to make them differs somewhat, mostly in the colors of the lining – blue, white or ivory. Some private hatters were French; the lining of these caps was black, the same color as for the French Marine Nationale.

▲ Kriegsmarine Stabsfeldwebel wearing the white summer cap which was worn from April 20 to September 30 while in Germany. Abroad it was worn according to the climate.

▲ Unusual silk lining that was originally yellow. The manufacturer's mark is stamped on the celluloid sweatshield. (Collection P.M Rousseau)

▲ Mannschaften/Maate summer Schirmmütze, with removable crown cover, black mohair band, embroidered nylon oak-leaf wreath, gold metal eagle. The lining is yellow silk and the sweatband is brown leather.

▶ Petty Officer and Officer-candidate's cap. This model has light alloy stamped oak-leaf wreath and eagle. (Collection P-M. Rousseau)

PETTY OFFICERS AND OFFICER CANDIDATES

▲ **Junior Officer's cap.** The navy blue crown cover is removable, the band is black mohair, the insignia and the visor's edge are embroidered in gold thread, and the cockade is woven. The lining is black. This cap has all the signs of having been worn for a long time. Size 56½.

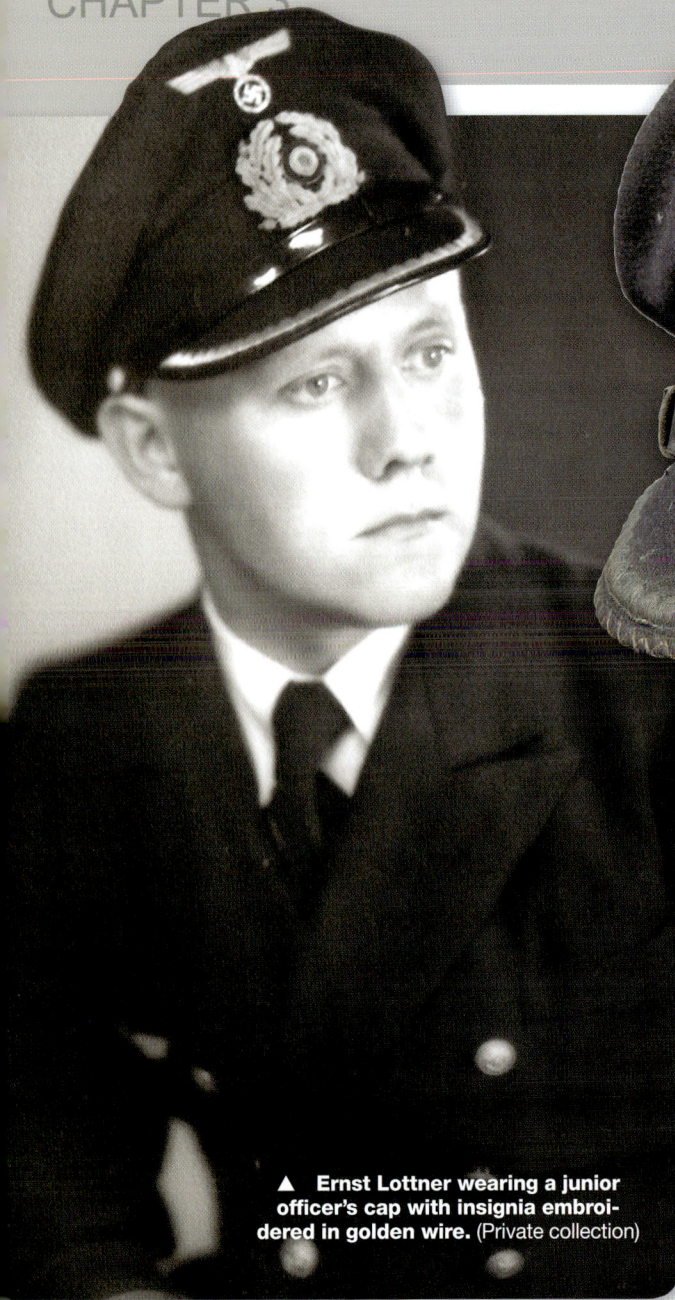

▲ Ernst Lottner wearing a junior officer's cap with insignia embroidered in golden wire. (Private collection)

▲ **Front and back of officer's cockades embroidered with gold wire (left) and in stamped light alloy (right).** (Collection P-M Rousseau)

▲ **Red silk lining in a rare model.** (Private collection)

▲ **Junior Officer's Navy-blue cap** belonging to a Leutnant zur See, Overleutnant zur See, or Kapitänleutnant with the insignia of the 1st U-Flotilla. The insignia is embroidered in gold wire.

▲ Eric Topp, legendary commander of U-552, in 1941. In this signed photo the Topp is wearing a white crowned cap, whose national emblem is a type usually reserved for the white or khaki tunic; a small whimsy accorded to this national hero. (Collection L. Berrafato, donated by Erich Topp in 1990)

▲ An Oberleutnant zur See poses for a portrait before he departs on a patrol during the winter of 1944/45. This young U-boat commander wears a white-crowned cap, which was authorized for U-boat commanders. He has been decorated with a U-boat Combat Clasp that appears to be silver. We can also see two other badges, one for U-boats, the other for minesweepers, showing that our officer once served on a surface ship. This superb photo allows us to admire a naval officer during the last months of the war. (Collection Laurent Berrafato)

▲ **Kriegsmarine Junior Officer's cap** with a white summer removable crown cover. The eagle is golden aluminum, the cockade and oak-leaf wreath are embroidered golden wire. The owner has pinned one of his initials on the insider leather edging and his surname on a visitor's card slipped into the allotted slot in the celluloid sweatshield. (Private collection)

▲ Victory pennants (Erfolgswimpel) were raised on every U-boat returning victorious from a patrol. The number 2500 indicates tonnage sunk, and in this case the ship type: a tanker. (Collection P-M Rousseau)

▶ U-boat commander, decorated with the Knight's Cross, sitting on one of his ship's guns, in front of victory pennants indicating tonnage sunk.

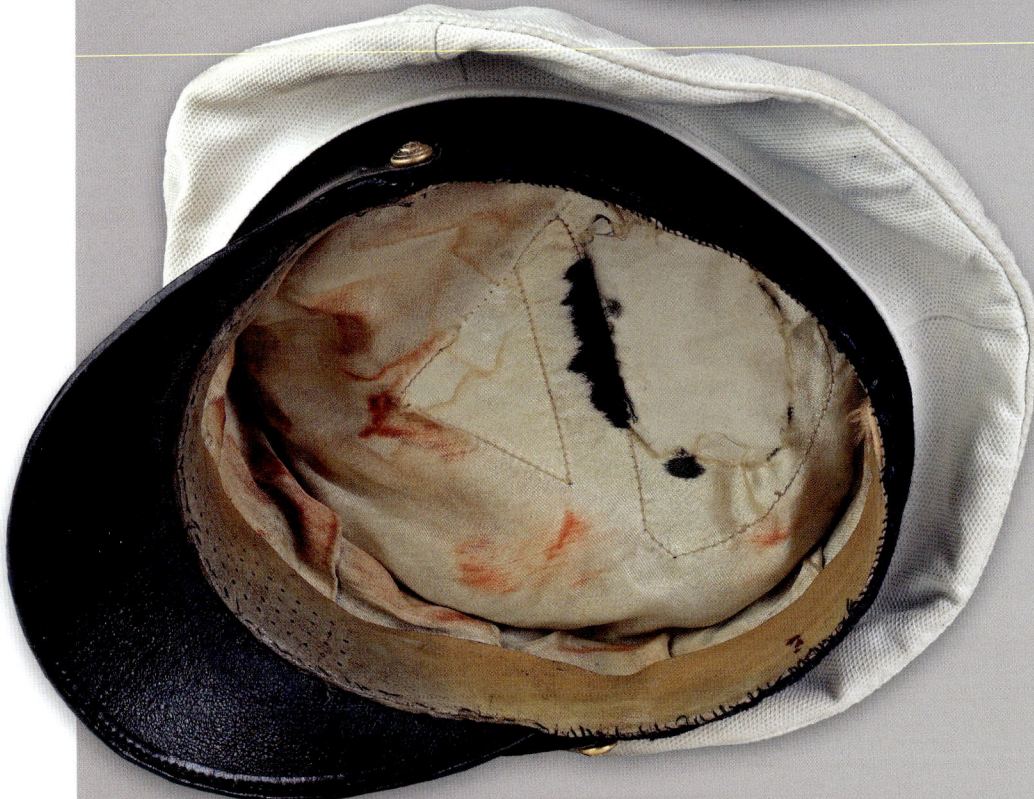

▲ April 1942, a U-boat crew returning from a patrol. We can see the white summer crown cover worn only by the commander of a U-boat, a distinctive sign; this cover is removable for easy cleaning and has a gold metal eagle.

▲ **Kriegsmarine Junior Officer's cap.** Removable white crown cover, metal eagle, mohair band, false leather chinstrap with gold buttons stamped with an anchor, visor embroidered in golden wire, oak-leaf wreath and cockade embroidered in silvered wire, silk lining, the cap's stiffener is at the back. This cap was found during the liberation of Bordeaux.

▲ **Navy-blue Junior Officer's Schirmmütze.** The numerous German hatters catering for the military didn't always use the same material and many of the caps on the market are of different material. (Collection L. Berrafato)

◄ **A Kapitän zur See (left) and a Korvettenkapitän (right) wearing navy blue senior officer's caps.** (Collection L. Berrafato)

SENIOR OFFICERS

◄ **Kriegsmarine Senior Officer's cap.** Blue crown with mohair band, eagle and oak-leaves embroidered in golden wire, and a large row of oak-leaves around the rim. The lining is blue silk.

▲ **Generaladmiral Alfred Saalwächter** (1883-1945), who commanded the Kriegsmarine in France, and decorated with the Knight's Cross; he was shot by the Russians in 1945 in Moscow, then reinstated by them in 1994. The cap was recovered by a member of the FFA and exchanged for two cartons of cigarettes in Berlin immediately after the war. This cap has a removable white crown cover. The national emblem is golden wire embroidered on a white background in the place where a metal eagle should be. The lining is yellow satin. This magnificent Schirmmütze was undoubtedly made before the war. (Collection XA, photo F. Coune)

▲ Generaladmiral Alfred Saalwächter is shown here wearing a blue cap with the M1935 national emblem, embroidered with silvered wire.

▶ We have decided to illustrate the Kriegsmarine through the legendary German **U-Boot-Waffe.** At left is a figure of an officer of the U-Boot-Waffe wearing a civilian checked shirt under his navy blue tunic surmounted by a fine blue officer's cap. The insignia is embroidered with golden wire, and the insignia of the 1st U-Flotilla, based in Brest, is pinned to the mohair band. The figure at right is that of a U-boat commander, who stands out from the rest of the crew because of his traditional cap with removable white crown cover. The eagle is golden aluminum, and the visor is trimmed with golden wire. Our commander wears a Heer tunic that most submarine crewmembers wore (this example is strictly regulation German, and a copy of British battledress) to which he has added shoulderboards indicating rank. He also wears the national emblem pinned above his right pocket, and on the left pocket is the U-Boat Combat Badge, awarded for two complete war patrols. Around his neck are binoculars enclosed in rubber, anti-shock casing. At bottom, on the flag of the Großadmiral (blue edging), are a number of items from the Kriegsmarine: an officer's dagger, a white shore patrol summer cap, a pair of shoulderboards belonging to a junior officer, a propaganda book about armed submarines, a Kriegsmarine officer's forage cap, a blue enlisted forage cap with the Normandy based minelayer unit's insignia, a pair of leather work gloves intended for U-boat crews, and a plate (manufactured by Henriot Quimper) from the Lanveoc Poulimic Base, and a tile (from the same supplier) with the insignia of the U-Flotilla based at Lorient, France. (Private collection, photo Frederic Coune)

ADMIRALS

▶ **Schirmmütze belonging to a Verwaltungsbeamte im Offiziersrang** (Administrative official accorded a Kriegsmarine officer's rank). The removable crown cover is of blue serge with a mohair band, silvered wire embroidered insignia, silver braid. There is no lining, and the sweatband is grey leather. Size 55.

▲ Navigation instruction cruise on the "Paul Beneke" around 1940. Instructor Kurz is wearing a blue Schirmmütze intended for administrative officials accorded a naval officer's rank. (Private collection)

▲ **Military official** given the rank of Navy Junior Officer, wearing a cap with a blue crown. (Collection L. Berrafato)

▲ **Cap with a white crown cover intended for a Military Official** given the rank of Kriegsmarine officer. The insignia is silver and the braid is silvered wire. (Collection L. Berrafato)

▶ **This Reichsmarine cap,** worn when in field uniform, is made of linen canvas. The headband and the piping are Rescda color, the peak is patent black and the insignia are metal. The cotton lining bears the manufacturer's name: "Bensel Oldenburg." This cap is very rare.

▲ This photo, taken in 1935, shows a 1921 model field cap with a field gray peak, worn by Reichsmarine officers when in field uniform. It was also worn during 1935 before being replaced by the new caps made for the Kriegsmarine. (Collection L. Berrafato)

▲ Taken the same day as the above photo, this snapshot shows an NCO wearing the new Schirmmütze made especially for the Kriegsmarine. (Collection L. Berrafato)

▶ Oak-leaf wreath embroidered in gilded wire, eagle and mohair headband from a Reichsmarine cap worn during the Reichswehr period.

▲ This interesting photo shows an NCO wearing an Alter Art cap that was usually only found in the Heer, but in this case the insignia is woven, in golden wire. (Collection L. Berrafato)

▲ Troop cap with woven oak-leaf wreath, metal eagle and green piping, false leather chinstrap with anchor buttons. Stamped on the inner celluloid sweatshield, "Wasserdicht" ("waterproof"). This type of cap usually has a blue satin lining.

▲ This sailor in a field-grey uniform is wearing a standard issue Schirmmütze that was given to officers and sailors. However, the eagle on the crown is not a regulation model. It looks more like the NSDAP symbol than an earlier model dating from 1934 to 1935. (Collection L. Berrafato)

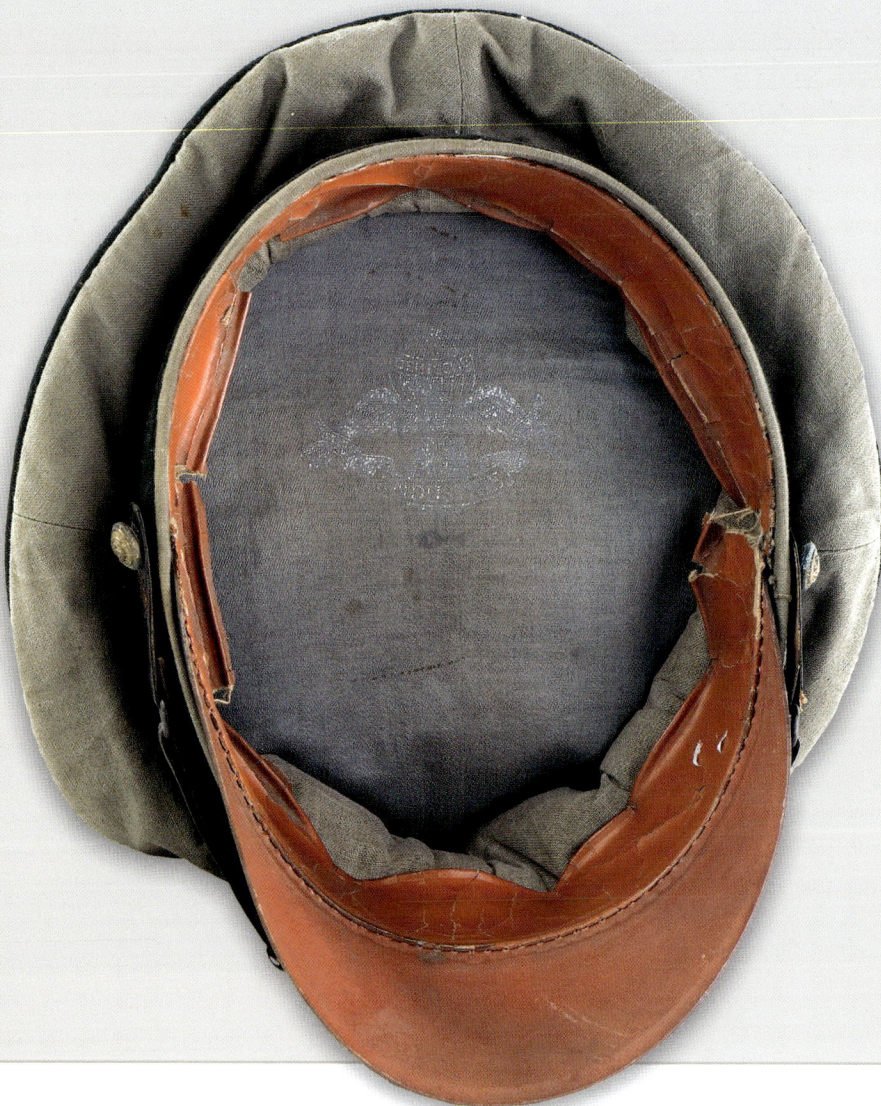

▲ **Kriegsmarine Officer's Field-Grey Schirmmütze.** This is the early model cap (pre-1935) with a dark, field green band and piping. The national emblem is metal rather than embroidered with silvered wire as on later models. The interior also shows that this is an early model: lining in cream colored satin (instead of blue), diamond shaped celluloid sweatshield (later models had a rounded end) with the manufacturer's stamp. (Photo F. Coune)

▲ **Military Official's Schirmmütze** (civil servant given Kriegsmarine officer's rank but taking a Heer training course). The insignia is embroidered in silvered wire (instead of gold insignia). (Private collection. Photo F. Coune)

▲ **Junior Kriegsmarine Officer** in field-grey uniform. His cap insignia is embroidered in golden wire. In 1935 the field-grey band and piping were changed to dark green to match those of the Heer. (Collection L. Berrafato)

CHAPTER 4

WAFFEN-SS

▶ SS-Sturmmann of the Leibstandarte Adolf Hitler, taken at Christmas 1940. He has a saddle-form cap with the eagle curved to fit the shape of the crown front. He has LAH insignia on his shoulderboards. (Collection Cyril Monties)

The **Waffen-SS** existed between 1940 and 1945 (prior to that, the unit was called the Verfügungstruppe). Though on the operational side it answered to the Wehrmacht, it was not one of its units, but belonged to the Schutzstaffel (SS) as a separate fighting force. As a fully independent unit, they were distinguished, not only by their intrinsic organization, but also by their unique uniform. One of the distinguishing signs that set the Waffen-SS apart from other units was undoubtedly the Totenkopf (death's head) insignia under which they fought. Taken from the German tank forces, its history is much older and dates back to the Prussian Hussars who fought before and during the First

World War and after them the Freikorps, and then to the first Nazi led troops. The SS designed their own, more realistic version of the Totenkopf insignia. The SS used this insignia as propaganda, reminding the German people of past noble armies and therefore attempting to give themselves a certain legitimacy. The SS wore the Totenkopf on the front of all their caps, including the Schirmmütze.

▲ Hauptscharführer "Spiess" of the SS Totenkopf Division. His cap has a Heer chinstrap, and the Totenkopf insignia is affixed over the piping.

▲ **Waffen-SS Infantry NCO and Troop Schirmmütze.** Serge with a black band and white piping. The lining is yellow rayon. The insignia is aluminum and the chinstrap is patent leather. Size 54½.

▲ **Front and back details of the national emblem.**

▶ This NCO, part of the Heimwehr Danzig Regiment, is wearing a cap with its characteristic Waffen-SS chinstrap.

NON-COMMISSIONED OFFICERS AND TROOPS

▲ **Waffen-SS Infantry NCO's cap.** The crown is field-grey, the band black, the piping white, the insignia is metal. It has a flat false chinstrap made of black leather. The lining is black.

▲ This Unterscharführer, decorated with the Iron Cross 2nd Class, the Infantry Assault Badge, and Sports Badge, is wearing his Schirmmütze pinched at the front, giving it a more aesthetic appearance.

▲ **Waffen-SS Signals NCOs and Troops cap.** Field-grey felt, black band, yellow piping, metal insignia and two-part leather chin strap. The lining is golden yellow rayon with a leather sweatband. Size 55½.

▲ **This Rottenführer is a member of the LAH as seen from his shoulderboards and sleeveband. He has slightly modified his cap: the chinstrap and stiffener have been removed and the top has been given a pinched shape.** (Private collection)

▲ **Feldmütze, 1938.** The crown and visor are field-grey serge, the eagle is woven but looks like one intended for the left sleeve of the SS uniform. The Totenkopf is stamped metal. At first, worn by the troops and NCOs, it was not unusual to find that a lot of officers preferred it.

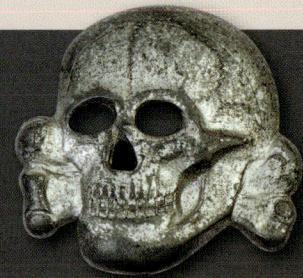

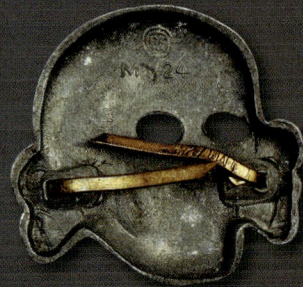

▲ **Fasteners on Waffen-SS Totenkopf insignia. Note the RZM marks.** (Private collection).

▲SS Division "Totenkopf" NCO wearing a cap whose chinstrap and stiffener have been removed giving it a "combat" look.

▲ This officer from a Waffen-SS panzer unit, a veteran of the 1st Winter on the Russian Front, is wearing a standard issue cap, whose stiffener has been removed. Note that the Totenkopf is fixed too high on the band.

▲ This Rottenführer, decorated with the Iron Cross 2nd Class, the 1st Winter on the Russian Front medal, the Infantry Assault Badge and the Wound Badge in Silver, is wearing a Schirmmütze with a very high crown.

▲ This much-decorated NCO (Infantry Assault Badge, Paratrooper Badge, and Wound Badge), is wearing a Schirmmütze that has been modified to an Alter Art Feldmütze (chinstrap and stiffener removed).

▲ SS-Obersturmbannführer Rolf Dalquen (whose brother Günter was the chief editor of the weekly "Schwarze Korps" and "SS-Standarte Kurt Eggers" commanding officer) is wearing a cap whose shape resembles an Alter Art Feldmütze, in spite of its vulcanized rigid visor. In fact he has removed the silver braid and the metal insignia that are normally found on a Schirmmütze and replaced them with woven versions.

▶ **The Waffen-SS**

On the figure at right, we have a tailor-made tunic belonging to a "Der Führer" Regiment officer (officer's armband embroidered in aluminum thread) and officer's insignia: eagle and collar insignia embroidered in silvered wire. He wears a leather belt with a round buckle – worn by Waffen-SS officers – and an officer's Schirmmütze with white piping (regulation color) for officers, but rarely followed. Our officer has taken care to pinch the top of the crown to break its standard form (a common practice at the time as seen in numerous period photos). On the figure at left is a model 1933 tunic modified in 1936 and worn by NCOs in the Leibstandarte Adolf Hitler. This time the insignia is embroidered with grey thread for NCOs and troops. He wears an M43 Feldmütze on which he has attached a Totenkopf in woven aluminum thread, which wasn't regulation but sometimes worn by NCOs as a way of standing out. In front, on the left: an M35 steel helmet with the SS double insignia from the "Das Reich" Division, a war trophy from an FFI fighter after the battle in the Dordogne during the 2nd SS panzer Division's advance to Normandy in June 1944; an enameled plaque from a field hospital, SS Lazaret Mogilev (from Jogny – Mogilev is a city in Belarus), Waffen-SS overseas cap, "Nordland" Division woven sleeveband, SS troop collar tab, "Totenkopf" Division troop collar tab, "Westland" arm insignia, Waffen-SS buckle (field-grey, factory issue) made by RODO. Panzer division (pink) officer's shoulderboard, another of an SS-Feldgendarmerie officer (orange) and a shoulderboard of a Waffen-SS infantry NCO (white piping), officer's sword knot, Soldbuch, identity tag, Wehrpass, and the sleeveband of a soldier in the SS-Postschutz. Lastly, a model 1943 Waffen-SS cap, a rarity as it has kept the regulation white piping for officers – a practice rarely followed as most officers preferred woven aluminum.
(Private collection. Photo Frederic Coune)

▲ **Details of different piping materials.**

▲ **Waffen-SS Infantry Commander's Schirmmütze.** The main characteristics of this cap are the fine gabardine, the black velvet band and the white piping. The eagle is made of fine zinc, the death's head has been added and the visor re-stitched. The lining is linen and the sweatband is grey ersatz material.

▲ Posing with his family, this SS-Obersturm-bannführer has been decorated with the Knight's Cross along with, among others, the Dienstauszeichnungder NSDAP Medal (award for long service in the Nazi Party). His Schirmmütze has a good "saddle form." The insignia is stamped aluminum and the silver braid is placed high on the band. (Collection Cyril Monties)

▲ The Polish Campaign. This SS-Unter-sturmführer, commander of the 11th Company, has just been decorated with the Iron Cross 2nd Class. He is holding a "Front hound," a dog that was perhaps his battalion's mascot at the front. His field tunic is sans collar tabs. His cap has a silver braid, a Heer eagle and a Totenkopf insignia set rather high, not in the center of the band and nearly touching the piping. The visor looks as if it is made out of vulcanized fiber rather than leather.

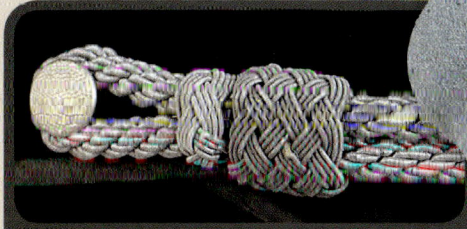

▲ Detail of a chinstrap button.

▲ **SS-Obergruppenführer Karl Wolff's personal cap.** Field-grey serge, black velvet band, silver piping, yellow silk lining with "Aug.Müller München" stamped inside, silver insignia. Wolff joined the NSDAP and the SS in 1931. From 1933 to 1943 he was Himmler's chief-of-staff, and liaison to Hitler; he was promoted to Gruppenführer und Generalleutnant der Waffen-SS in 1942, and then became Military Governor of Northern Italy, and Höchste SS- und Polizeiführer and military commander in Italy to the end of the war. Size 57.

◄ **Sepp Dietrich in Obergruppenführer und General der Waffen-SS uniform.** In this portrait he wears the Knight's Cross with Oak Leaves.

DEINE ZUKUNFT

▲ The cover of a propaganda booklet with a drawing by Anton and presenting "Your Future" to young members of the Reich, showing an SS Untersturmführer wearing a superb "saddle form" Schirmmütze with regulation metal insignia.

▲ **Field-Marshal Graziani and SS-General Wolff inspecting the troop of an SS Polizei Battalion in the winter of 1944.**

CAP PIPING OF THE HEER

▲ Infanterie
Leutnant

▲ Motorcycle Troops
Oberfeldwebel

▲ Artillerie
Oberstleutnant

▲ Nebeltruppe
Leutnant

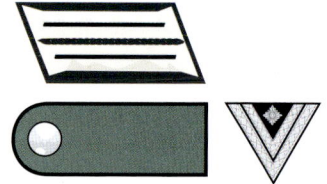

▲ Pionier
Stabsgefreiter

▲ Transportation
Feldwebel

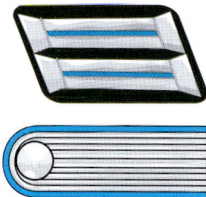

▲ Transportation
Leutnant

▲ Pionier and
Feldgendarmerie
Major

▲ Judicial
Oberstleutnant

▲ Panzergrenadier
Unterfeldwebel

▲ Gebirgsjäger
Unteroffizier

▲ Kavallerie
Hauptmann

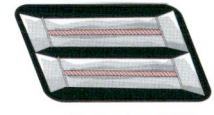

▲ Panzer
Oberst

▲ Pionier
1939-1943
Stabsfeldwebel

▲ Signals
Oberleutnant

▲ Medical
Hauptmann

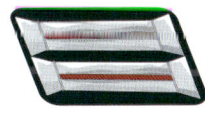

▲ Veterinary
Leutnant

▲ Official
Oberleutnant

▲ Propaganda
Obergefreiter

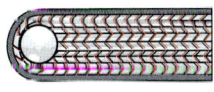

▲ Sonderführer

▲ Chaplain
Heerespfarrer,
Heeresoberpfarrer

Infographies, André Jouineau© Uniformes 2010

CAP PIPING OF THE LUFTWAFFE

▶ Gold: General Officers		◀ Silver: Officers	
▶ Yellow: Navigators and Paratroopers		◀ Red: Flak	
▶ Brown: Signals		◀ Dark Blue: Medical	
▶ White: Hermann Göring Division		◀ Dark Green: Luftwaffe Administration	
▶ Light Green: Air Traffic Control		◀ Rose-pink: Technical and Engineer	
▶ Black: Luftwaffe Construction Units		◀ Light Blue: Luftwaffe Reserve	

Cap with Veterinary piping.

Infographie © Uniformes 2010

CAP PIPING OF THE WAFFEN-SS

▶ Silver: General Officers		◀ White: Staff, Infanterie and Grenadiers	
▶ Lemon Yellow: Signals and War Correspondents		◀ Golden Yellow: Kavallerie and Reconnaissance	
▶ Rose-Pink: Panzer		◀ Dark Blue: Medical	
▶ Black: Pioniers		◀ Red : Artillerie	
▶ Light Green: SS Polizei and Mountain Units		◀ Orange : Feldgendarmerie	
▶ Orange Red: Replacement Units		◀ Crimson: Veterinary	
▶ Bordeaux Red: Judicial		◀ Dark Red: Nebeltruppe	
▶ Pale Pink: Transport and Maintenance		◀ Light Blue: Administration	
▶ Dark Green: Reserve Officers		◀ Black and White: Armored Engineers, Concentration Camp Guards	

Infographie © Uniformes 2010

ELEMENTS OF A CAP

The crown is made of field-grey knitted material (Trokost-off) woven in parallels (grey-blue for the Luftwaffe and blue for the Kriegsmarine). Superior quality caps intended for officers were often made from a combed wool-based textile called "Eskimo." Usually, however, officer's caps were made from a wool-like textile called "Döskin."

❶ National emblem (Hoteitsabzeichen): from 1934 the traditional cockade in state colors was no longer in use and was replaced by a spread-winged eagle carrying a swastika: the national emblem, like the oak-leaf wreath, was first made of nickel silver, then of a lighter alloy with an aluminum base, and zinc later in the war. In 1935, along with the oak-leaf wreath, an eagle with longer wings was introduced.

❷ Piping (Paspelierung): made of different materials, mainly rayon and serge, and mounted on different supports: wood, cord, etc. Piping was often about 3mm wide on standard issue caps, but sometimes much larger on privately purchased caps, notably those belonging to officers.

❸ First pattern oak-leaf wreath (Eichenlaubkranz): was adopted shortly after World War I and was used until a new model was introduced in 1935. Insignia embroidered in silvered wire was permitted on privately purchased caps until 1942 (at first they had a field-grey background, and then dark green). The new wreath, designed by Paul Casberg, was at first made of nickel silver, then out of an aluminum based alloy, and finally of zinc (duller, greyish).

❹ Cockade (Reichskokarde): from 1933, the gold oval with the Reichswehr eagle went out of fashion and was replaced by a round cockade in the national colors. Different versions were available and often the red center was made of felt. The oak-leaf wreath embroidered in silvered wire surrounding the metal cockade appeared and was worn concurrently.

❺ Band (Besatzstreifen): at first this was field-grey and then changed to dark green. A bluish color was the regulation for the Army during World War II. The Luftwaffe and Kriegsmarine band was made of black mohair. Officer's braid (Kordel) was made from silvered wire and later from aluminum. Officer candidates were also allowed to wear them. General officers had gold braids.

❻ Chinstrap (Sturmriemen): for NCOs and the troops, this was usually made of blackened varnished leather or some sort of ersatz material, fixed by black varnished buttons. The buckles in various forms were metal and also varnished in black (round buckles were typical in the Luftwaffe); the Kriegsmarine (for all ranks, except for the administrative officials) wore chinstraps comprising two leather bands – and two loops – fixed to each other by a snap fastener, and to the band by two golden buttons.

❼ Visor (Mützenschirm): was made of vulcanized fiber and varnished black.

Cap Makers

EREL

Robert Lubstein (EREL-Sonderklasse, the phonetic sound of Lubstein's initials: RL) was founded in 1902 in Berlin, NW21, Alt Moabit 105, only for cap making. He changed his address in 1939: Berlin No.55, Heinrich-Roller-Strasse 16-17 (also after the war). In 1948 Margarete Lubstein (his wife) took over the company. EREL patented a modified sweatband with an air cushion to relieve the visor's pressure on the forehead in the summer of 1940. EREL was the official supplier to the Heer's uniform store (Heeres Kleiderkasse), but also supplied all other branches, including the Waffen-SS. Until 1939 Lubstein had two separate branches for selling his caps: one under his full name and the other with the phonetic initials, EREL-Sonderklasse. EREL was the biggest supplier of uniform caps in Berlin and their publicity slogan was, "The most modern and elegant officer's caps." Different classes and qualities of caps were also available, from the simplest to the most elaborate: 140, Standard, Extra, Privat. Apart from their prestigious Schirmmütze, EREL also produced Shakos, parade caps (made of fiber), and tropical caps. Several distributors stocked the well-known brand: Militäreffekten Sperrling in Dresden, Militärwarenhaus Hans Dürbeck in Vienna (Wien IX, Berggasse 31), Wilhelm Voigt in Magdeburg, Thiele and Edellinger in Darmstadt, and finally V. Osterwalder in Lüneburg. Their names appeared above the diamond EREL logo on the celluloid sweatshield.

Peküro

Peter Küpper (Peküro = Pe (ter) Kü (pper) Ro (nsdorf) – the Wuppertal Quarter-North-Rhine, Westphalia, where he had his business). His address was Krim 30, Wuppertal-Ronsdorf. The firm was founded around 1894 and disappeared recently. The name Peküro would seem to have been first used in 1935. The owners were Peter Küpper (died in 1934), Berhardine Zimmerman (Head Accountant), Heinz Joachim Zimmerman-Baum and Claudia Kolter (nee Zimmerman-Baum). The logo was a diamond with the inscription "Peküro" in the center, which was found on the celluloid sweatshield. Peküro manufactured high quality peaked caps for the Heer, Luftwaffe, police, fire brigade, as well as standard issue serge caps and Bergmützen. Peter Küpper also had a patent (taken out in 1931) for a forehead protection feature. His distributors were based in Aix-la-Chapelle and in Offenbach (Hermann Schellhorn). Like EREL, Peküro didn't hesitate calling itself, "the biggest manufacturer of caps in Germany," and to use the logo, "Peküro represents quality and progress." After the war Peküro supplied caps to the Bundesgrenzschutz (Federal Border Guard), after Carl Halfar was forced to close down. Anecdote: Peter Küpper's firm later sponsored the Winter Olympic Games and specialized in sports caps under the name of CODEBA and had a site on the internet. One of Peküro's particularities was a thin satin or velvet band around the edge of the visor at its junction with the inside sweatband and passing under the latter.

Schellenberg

August Schellenberg's famous "Bear" logo, Uniformmützen-Fabrik, Berlin O 27, Alexanderstrasse 40. This firm was created in the beginning of the 1930s and lasted until the mid-1950s. Even though the logo was usually red, in some models it was black or silver. Schellenberg made caps for the Allgemeine-SS, NSDAP, Heer, and Lufwaffe. During the mid-1930s his production was estimated at 300 caps a month. At the end of 1936 the firm was awarded with the DRGM (Deutsches Reich Gebrauchmuster) protected by the law (or gesetzlich geschutzt, which is often found in abbreviated form) for a feature that was intended to prevent pressure on the forehead. These days the only manufacturer from World War II still in existence is Albert Kempf (Alkero) & Co. GmbH. Like Peküro, it started out in Wuppertal and was a competitor before finally overtaking them.

◀▲ A variety of maker's marks found on inner sweatbands.

◀▲ Peküro maker's marks.

OTHER MANUFACTURERS

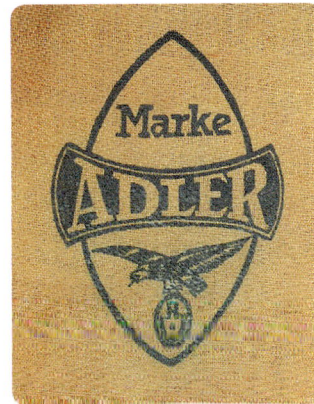

▲ Various maker's marks on sweatband interiors.

SOVIET CAPS MADE IN GERMANY

◄ **Infantry cap, Model 1935. Made by Robert Lubstein-EREL, 1948.** (Collection and Photo Richie C., via Philippe Rio)

► **Infantry cap, Model 1935. Made by Robert Lubstein-EREL, 1949.** (Collection and Photo Richie C., via Philippe Rio)

► **Tank unit cap, Model 1935. Manufacturer unknown.** (Collection and Photo Richie C., via Philippe Rio)

◄ **Tank unit cap, Model 1935. Manufactured c.1946-50.** (Collection and Photo Richie C., via Philippe Rio)

◄ **Flight Personnel Officer's cap, Model 1949. This example was intended for an officer's dress uniform. Made by Robert Lubstein-EREL, 1950.** (Collection and Photo Richie C., via Philippe Rio)

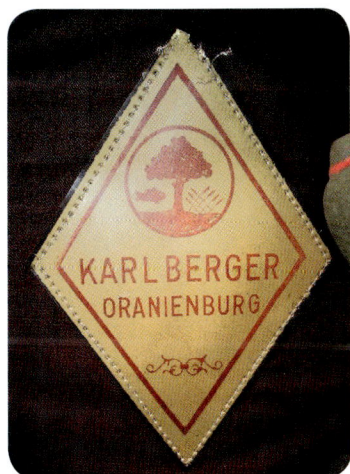

KARL BERGER
ORANIENBURG

► **Tank unit cap, Model 1935. Made by Karl Berger during the 1950s.** (Collection and Photo Richie C., via Philippe Rio)

ABOUT THE MARKINGS

On top: the manufacturer's brand name (Robert Lubstein for example), below, the town or city (Berlin) and in the square, the size, the quality and the year of fabrication. EREL's address was Berlin No.55, Heinrich-Roller-Strasse 16-17 since 1939 and also after the war.

The only differences between German and Soviet caps are that the German caps were better made. They have German manufacturer's markings and usually a leather sweatband (whereas in the Soviet caps this item was made of synthetic material). Otherwise, the cut, the band, the piping, material, etc. conforms to the Red Army's order form. This type of cap was manufactured until the creation of the DDR, doubtlessly at the beginning, as part of war reparations. Among other types of caps, you may also find German made forage caps worn by Soviet officers.

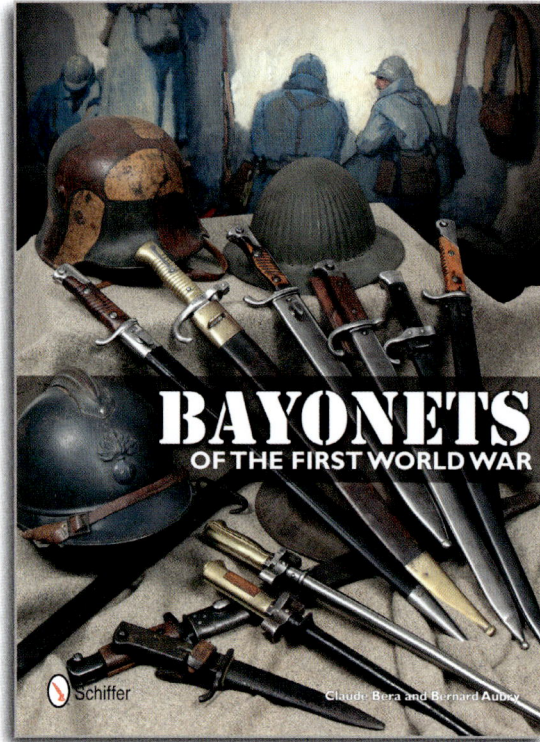

BAYONETS OF THE FIRST WORLD WAR

Claude Bera and Bernard Aubry.
This detailed, concise look at WWI bayonets presents their development and use by all combatant countries. Images show full views of bayonets and accessories, and a clear look at hilts, connection points, blades, and a wide variety of manufacturer's markings.

Size: 9" x 12" • 273 color photos • 80 pp.
ISBN: 978-0-7643-4459-6 • hard cover • $29.99

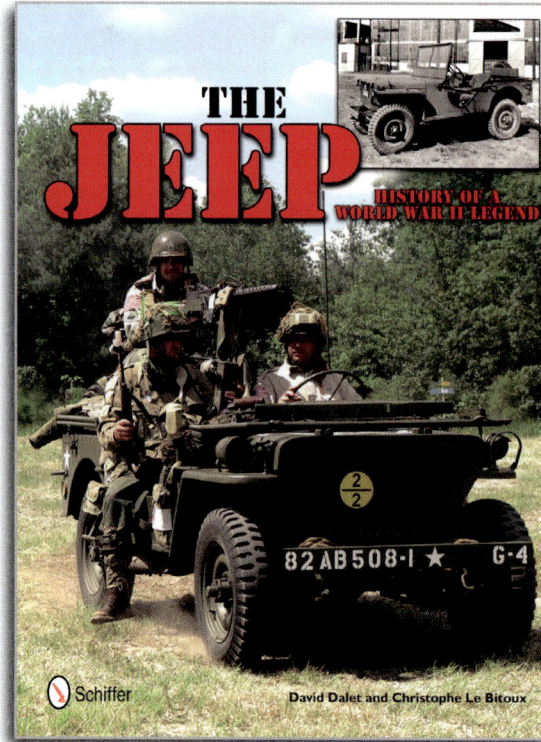

THE JEEP: HISTORY OF A WORLD WAR II LEGEND.

David Dalet and Christophe Le Bitoux. Nearly 200 high-quality color, and World War II era photos show the Jeep in its many uses, and up-close, detailed images show the differences between variant types. This book is a concise reference for the Jeep enthusiast, historian and restorer.

Size: 9" x 12" • 196 color photos • 80 pp.
ISBN: 978-0-7643-4460-2 • hard cover • $29.99

Schiffer books may be ordered from your local bookstore, or they may be ordered directly from the publisher by writing to:

Schiffer Publishing, Ltd.
4880 Lower Valley Rd.
Atglen, PA 19310
(610) 593-1777; Fax (610) 593-2002
E-mail: Info@schifferbooks.com

Please visit our website catalog at www.schifferbooks.com or write for a free catalog.

Printed in the United States of America